*An
International
Sourcebook
of
Paper History*

An
International
Sourcebook
of
Paper History

by Irving P. Leif

ARCHON
DAWSON
1978

© Irving Leif 1978

First published in 1978

Archon Books, The Shoe String Press, Inc.
995 Sherman Avenue, Hamden, Connecticut 06514 USA

Wm Dawson & Sons Ltd, Cannon House
Folkestone, Kent, England

Archon ISBN 0-208-01691-0
Dawson ISBN 0-7129-0827-7

Printed in the United States of America

Library of Congress Cataloging in Publication Data

Leif, Irving P.   1947-
        An international sourcebook of paper history.

        Includes indexes.
        1. Paper working and trade—History—Bibliography.
I. Title.
Z7914.P2L4   [TS1090]   016.676'.209   77-20485

Contents

INTRODUCTION

The invention of paper in China by Ts'ai Lun in 105 C.E. remains one of the most important events in history, and paper has had its most glorious moments in the age of books and in the dissemination of knowledge throughout the world. These moments have been studied and written about by many people from different countries and cultures. And yet, paper historians have been simultaneously swamped by material in various languages and handicapped by their lack of a thorough guide to the literature.

This book, then, is the first comprehensive bibliography on the history of handmade paper, from its earliest appearance to the manufacture of paper by machine. Pre-paper writing materials are noted too, insofar as their relationship to paper is concerned; therefore, the reader will find only selected articles on papyrus, vellum, parchment, and so on, listed as well.

In compiling the literature for this bibliography, a complete search was made of both English and non-English books and journals in libraries all over the United States. Every major journal that has published paper historical research was examined, as well as many obscure ones. To maintain scholarly exactness, all citations are as they appeared originally in print. While this can mean erratic forms in the non-English listings, it will save users time and effort if they are unable to obtain a book or journal but want to cite the work in their writing. To further overcome language barriers, the literature in each pertinent part of the bibliography has been listed by country.

Both a subject index and an author index are provided at the end of the book to help users find specific topics in paper history. Major subjects are entered along with detailed listings of subjects and events by country; the author index makes the work of individual paper historians accessible immediately.

It is the author's hope that this bibliography will be valuable to paper historians, book collectors, librarians, students of history, and others interested in this important aspect of man's intellectual development and history.

I would like to thank Pamela S. Bracken for her assistance in the preparation of this book.

IRVING P. LEIF
1978

1 General Histories
of
Papers and Watermarks

PAPER

1   Aiken, P. Henderson. 1914. "Some Notes on the History of Paper."
Transactions of the Bibliographical Society (London)
13 (November): 201-217.

2   Alibaux, Henri. 1939. "L'Invention du Papier." *Gutenberg-Jahrbuch*
14: 9-30.

3   _____. 1921. "Les Légendes de L'Histoire du Papier, les Moulins
à Papier sur L'Hérault en 1189, le Papier de Coton." *Revue du
Lyonnais* No. 3 (July-September): 345-370.

4   _____. 1939. "Moderne Papyrusherstellung in Syrakus."
Wochenblatt für Papierfabrikation 70 (March 18): 241-244.

5   Arvay, Wiktor. 1955. "Najstarszy Znany Regulamin Papierni."
Przegląd Papierniczy 11 (January): 12.

6   Audin, Marius. 1928. "De L'Origine du Papier Vélin." *Gutenberg-
Jahrbuch* 3: 69-86.

7   _____. 1929. "De L'Origine du Papier Vélin." *Gutenberg-
Jahrbuch* 4: 312.

8   Babinger, F. 1931. "Papierhandel und Papierbereitung in der
Levante." *Wocheblatt für Papierfabrikation* 62 (December):
1215-1217.

9   Bacquie, F. 1926. "Les Progrès de la Papierterie au XVIIIme
Siècle." *Le Papier* 29: 683-690.

10 Basanoff, Anne. 1965. *Itinerario Della Carta: Dall'Oriente
All'Occidente e Sua Diffusione in Europa.* Milan, Italy:
Edizioni Il Polifilo.

11 Bayley, H. 1908. "The Romance of Papermaking." *Paper Making*
27 (May): 175, 177-179.

12 Beadle, Clayton. 1898. "The Recent History of Paper Making."
Journal of the Royal Society of Arts 46: 405-416.

13 _____, and Henry P. Stevens. 1909. "Hand-Made Papers of
Different Periods." *Journal of the Royal Society of Arts* 57
(February 26): 293-304.

14 Belani, E. 1940. "Einer Der Ältesten Kollergang-Typen."
Wochenblatt für Papierfabrikation 71 (August 17): 414.

15 _____. 1924. "Zur Geschichte des Banknotenpapiers."
Wochenblatt für Papierfabrikation 55 (October 25): 2744.

16 _____. 1925. "Zur Geschichte der Papiermacherkunst." *Der
Papier-Fabrikant* 23 (March 22): 193-194.

17 Berthelé, Joseph. 1906. "Un Prétendu Moulin à Papier sur l'Hérault
en 1189." *Le Bibliographe Moderne* 10: 201-213.

18 Blades, William. 1889. "On Paper And Paper-Marks." *The Library (First Series)* 1: 217-223.

19 Blanchet, Auguste. 1900. *Essai sur l'Histoire du Papier et de Sa Fabrication.* Paris, France: Leroux.

20 Blücher, Gebhard, 1969. "Kronstadter Drucke und Papier des 16. Jahrhunderts." *Gutenberg-Jahrbuch* 44: 11-17.

21 Blum, André. 1934. *On the Origin of Paper.* New York: R. R. Bowker.

22 _____. 1932. "Les Origines du Papier." *Revue Historique* 170: 435-447.

23 _____. 1946. *Le Route du Papier.* Grenoble, France: Éditions de l'Industrie Papetière.

24 Bockwitz, Hans H. 1938. "Die Ältesten Abbildungen zur Geschichte des Pergamenter-Handwerks." *Wochenblatt für Papierfabrikation* 69 (December 3): 1017-1020.

25 _____. 1950. "Der Erste Grossauftrag auf Papier. Historische Betrachtung." *Druckspiegel* 5: 51-56.

26 _____. 1938. "Die Früheste Erwähnung und Älteste Abbildung des 'Holländers'." *Wochenblatt für Papierfabrikation* 69 (October 8): 840-842.

27 _____. 1939. "Zur Geschichte des Papiers und Seiner Wasserzeichen." *Archiv für Buchgewerbe und Gebrauchsgraphik* 76 (August): 417-440.

28 _____. 1940. "Die Letzte Windpapiermühle." *Archiv für Buchgewerbe und Gebrauchsgraphik* 77 (January): 29.

29 _____. 1942. "Mittenwald als Durchgangsstation für Papier im Mittelalter." *Archiv für Buchgewerbe und Gebrauchsgraphik* 79: 496.

30 _____. 1936. "Neudruck Eines Papiergeschichtlichen Dokuments." *Archiv für Buchgewerbe und Gebrauchsgraphik* 73: 520.

31 _____. 1950. "Ostasien Denkt Anders Über das Papier als Wir." *Druckspiegel* 5: 448.

32 _____. 1949. "De Papiermaker." *De Papierwereld* 4 (July): 10-11.

33 _____. 1947. "Seit Wann Gibt Es Frauenarbeit im Papiermacher- und Druckgewerbe?" *Das Papier* 1 (August): 72-74.

34 _____. 1932. "Wie Sah das Älteste Papier aus?" *Archiv für Buchgewerbe und Gebrauchsgraphik* 69: 234.

35 Bogdán, István. 1969. "Einige Technische Daten zur Papiermacherei

im 17. bis 19. Jahrhunderts." *Papiergeschichte* 19 (November): 36-39.

36 _____. 1964. "Miszellen zur Papiergeschichte." *Papiergeschichte* 14 (December): 61-63.

37 _____. 1959. "Néhány Adat a XVI-XVIII. Századbeli Papírkészítésről." *Papíripar és Magyar Grafika* 3 (September-October): 189-194.

38 _____. 1955. "Papírgyártás 200 év Elött." *Papir-és Nyomdatechnika* 7: 263-267.

39 Buehler, J. Marshall. 1964. "A Paper Machine With a 'Hallmark'." *The Paper Maker* 33 (September): 45-50.

40 Bullock, Warren B. 1933. *The Romance of Paper.* Boston, Massachusetts: R.G. Badger.

41 Camac, C.N.B. 1931. "The Papyrus Industry of the Ancients." *Proceedings of the Charaka Club* 7: 72-81.

42 Carus, Gustave. 1933. "The Age of Paper." *Open Court* 47: 271-277.

43 Clapperton, R.H. 1954. "The Invention and Development of the Endless Wire, or Fourdrinier, Paper Machine." *The Paper Maker* 23 (February): 1-17.

44 _____. 1934. *Paper and Its Relationship to Books.* London, England: J.M. Dent and Sons.

45 _____. 1934. *Paper: an Historical Account of Its Making by Hand From the Earliest Times Down to the Present Day.* Oxford, England: Shakespeare Head Press.

46 _____. 1967. *The Paper-Making Machine. Its Invention, Evolution and Development.* Oxford, England: Pergamon Press.

47 Clemensson, Gustaf. 1953. *Papperets Historia Intill 1880.* Stockholm, Sweden: Hugo Gebers Förlag.

48 Cornely, Berthold. 1963. "Eine Papiergeschichtliche Untersuchung Über das Färben im Zeug." *Papiergeschichte* 13 (December): 53-62.

49 _____. 1956. "Eine Papiergeschichtliche Untersuchung Über das Schönen und Färben des Papieres in der Masse." *Papier geschichte* 6 (September): 49-60.

50 Deléon, Marcel. 1945. *Deux Grandes Figures de la Papeterie: Nicolas Robert, Inventeur de la Machine à Papier Continu et Aristide Bergès, Père de la 'Houille Blanche'.* Paris, France: Editions Elzévir.

51 Denne, Samuel. 1795. "Observations on Papers." *Archaeologia* 12: 114-131.

52   Devauchelle, R. 1958. "Le Papyrus." *Revue des Papiers* 1 (May): 23-24.

53   De Veer, B.W. 1953. "Ondergang van de Hollandse Papierfabrieken!?" *De Papierwereld* 7 (April): 279-280.

54   Diller, G. 1935. "Zünftiges aus Handwert und Papierhandwert vor 200 Jahren." *Wochenblatt für Papierfabrikation* 66 (December 21): 956-958.

55   Dvořák, F. 1958. "Výroba Ručniho Papiru." *Svĕt Techniky* 9: 302-307.

56   Elliott, Harrison. 1956. "Cartridge Paper: Its Background, Past and Present." *The Paper Maker* 25 (September): 11-12.

57   ————. 1954. "The Evolution of Newsprint Paper." *The Paper Maker* 23 (September): 45-49.

58   ————. 1954. "The Fourdrinier Wire: Its History, Structure, and Function." *The Paper Maker* 23 (February): 21-25.

59   Ernst, Gertrud. 1960. "Versippung von Papiermachergeschlechtern." *Papiergeschichte* 10 (February): 1-4.

60   Evans, Lewis. 1896. *Ancient Paper-Making.* London, England: Dickinson Institute.

61   Felix, D.A. 1952. "What Is the Oldest Dated Paper in Europe?" *Papiergeschichte* 2 (December): 73-75.

62   Filip, J.J. 1946. *Dĕjiny Papíru.* Prague, Czechoslovakia: Dílo.

63   Fiskaa, H.M. 1967. "Das Eindringen des Papiers in die Nordeuropäischen Länder im Mittelalter." *Papiergeschichte* 17 (June): 28-32.

64   ————. 1938. "Papirhistorisk Nyorientering i de Siste 50 År." *Papir-Journalen* 26 (November 30): 281-285.

65   ————. 1938. "Papirhistorisk Nyorientering i de Siste 50 År." *Papir-Journalen* 26 (December 15): 324-331.

66   Franklin, Benjamin. 1793. "Description of the Progress to be Observed in Making Large Sheets of Paper in the Chinese Manner, With One Smooth Surface." *Transactions of the American Philosophical Society* 3: 8-10.

67   Fritsch, Werner. 1974. "Über eine Sammlung Alter Buntpapiere." *Papiergeschichte* 24 (November): 16-23.

68   Füllner, Eugen. 1921. "Schlichte Gedichte." *Wochenblatt für Papierfabrikation* 52 (January 8): 26.

69   Gachet, Henri. 1954. "Des Premiers Papiers aux Premiers Filigranes." *Le Courrier Graphique* No. 72 (May): 27-36.

70   ————. 1953. "Réaumur, les Guêpes et le Papier." *Papiergeschichte* 3 (December): 75-79.

71 _____. 1938. "Six Siècles d'Histoire du Papier." *Le Courrier Graphique* No. 3: 3-9.

72 Gasparinetti, A.F. 1953. "Keine Verallgemeinerungen!" *Papiergeschichte* 3 (September): 48-49.

73 _____. 1957. "Zwei Alte Papiermühlen, die Nie Existiert Haben." *Papiergeschichte* 7 (April): 23-36.

74 Grand-Carteret, John. 1913. *Papeterie & Papetiers de l'Ancien Temps.* Paris, France: en l'Officine de Georges Putois, Maître Marchand-Papetier Ancien Juré et Garde de la Communauté.

75 Granniss, Ruth Shepard. 1935. "Some Recent Books About Paper." *Colophon* 1 (Summer): 133-142.

76 Gronow, W. Elsner von. 1925. "Kritische Betrachtungen Über die Verwendung von Baumwolle in der Papierfabrikation des Altertums und des Mittelalters." *Der Papier-Fabrikant* 23 (June 28): 416-419.

77 Grosse-Stoltenberg, Robert. 1972. "Papiermühlengeschichten." *Papiergeschichte* 22 (December): 46-59.

78 Grull, Georg. 1960. "Riesumschläge aus Oberösterreich." *Gutenberg-Jahrbuch* 35: 19-27.

79 Haemmerle, Albert. 1966. "Buntpapier. Geschichte und Techniken." *Schweizerisches Gutenbergmuseum* 52: 40-43.

80 _____. 1966. "Die Buntpapiersammlung Olga Hirsch." *Philobiblon (Hamburg)* 10 (June): 104-109.

81 _____. 1972. "Die Geschichte der Genji-Schriftrollen." *Papiergeschichte* 22 (December): 64-68.

82 _____. 1969. "Die 'Preisswürdige Papiermacheikunst' von 1699." *Papiergeschichte* 19 (December): 64-65.

83 _____. 1961-1962. "Wie das Buch 'Buntpapier' Entstanden Ist." *Imprimatur (New Series)* 3: 270-271.

84 _____, with Olga Hirsch. 1961. *Buntpapier.* Munich, Germany: Verlag George D.W. Callwey.

85 Hazen, Allen T. 1935. "Eighteenth-Century Quartos With Vertical Lines." *The Library (Fourth Series)* 16 (December): 337-342.

86 Heimann, W. 1928. "Die Grundstoffe der Papiererzeugung in den Jahrtausenden." *Wochenblatt für Papierfabrikation* 59 (September 15): 1027-1032.

87 Higham, Robert R.A. 1963. "The History of Papermaking." In *A Handbook of Papermaking,* by Robert R.A. Higham. London, England: Oxford University Press. 262-269.

88 Hunter, Dard. 1926. "About Hand-Made Papers and Deckle Edges." *The American Printer* 82 (April 20): 31-33.

89   _____. 1927. "About the Book, Primitive Papermaking." *The American Printer* 84 (December): 64-65.

90   _____. 1915. "Ancient Papermaking." *The Miscellany* 2 (December): 67-75.

91   _____. 1921. "A Bibliography of Marbled Papers." *Paper Trade Journal* 22 (April): 52-58.

92   _____. 1923. "The Couching Material Used By the Old Paper-makers." *Alfelco Facts* 2 (October): 3-7.

93   _____. 1924. "Features of Early Paper Making." *The American Printer* 79 (July 20): 27-30.

94   _____. 1926. "Fifteenth Century Papermaking." *Ars Typographica* 3 (July): 37-51.

95   _____. 1927. *Fifteenth Century Papermaking.* New York: Press of Ars Typographica.

96   _____. 1940. "Handmade Paper Moulds." *The Paper Industry* 21 (February): 1164-1169.

97   _____. 1921. "Laid and Wove." *Printing Art* 38 (September): 33-40.

98   _____. 1925. *The Literature of Papermaking, 1390-1800.* Chillicothe, Ohio: The Mountain House Press.

99   _____. 1923. *Old Papermaking.* Chillicothe, Ohio: The Mountain House Press.

100   _____. 1938. "Papermaking." In *A History of the Printed Book,* edited by Lawrence Wroth. New York: Limited Editions Club. 345-370.

101   _____. 1931. *Paper-Making in the Classroom.* Peoria, Illinois: The Manual Arts Press.

102   _____. 1947. *Papermaking: the History and Technique of an Ancient Craft.* New York: Alfred Knopf. 2nd Edition.

103   _____. 1930. *Papermaking Through Eighteen Centuries.* New York: W.E. Rudge.

104   _____. 1931. "Peregrinations & Prospects." *Colophon (New Graphic Series)* No. 7: 61-74.

105   _____. 1920. "A Rare Book on Papermaking." *Paper* 22 (September 15): 16-18.

106   _____. 1929. "The Romantic History of Paper Making. Rise of the Art in China and Its Development in Europe and the West." *The Paper Industry* 10 (March): 2160-2162.

107   _____. 1946. "Some Notes On Oriental and Occidental Paper and Books." *The Paper Maker* 15 (September): 3-13.

108 _____. 1937. "The Story of Paper." *The Natural History Magazine* 40: 577-597.

109 _____. 1950. "The World's Most Important Material—Paper—Its Origins and History." *The Illustrated London News* 217 (September 16): 437-440.

110 Ingpen, Roger. 1924. "Decorated Papers." *The Fleuron* No. 2: 99-102.

111 Irigoin, Jean. 1953. "Les Débuts de l'Emploi du Papier à Byzance." *Byzantinische Zeitschrift* 46: 314-319.

112 _____. 1972. "Quelques Innovations Techniques Dans la Fabrication du Papier: Problèmes de Datation et de Localisation." *Papiergeschichte* 22 (December): 59-64.

113 Jahans, Gordon A. 1934. "A Brief History of Paper." *Book Collector's Quarterly* 5 (July-September): 42-58.

114 Janata, G. 1914. "Aus des Handwerks Vergangenen." *Der Papier-Fabrikant* 12 (June 19A): 21-25.

115 Jarvis, Rupert C. 1959. "The Paper-Makers and the Excise in the Eighteenth Century." *The Library (Fifth Series)* 14 (June): 100-116.

116 Johnson, Robert K. 1953. *A Survey of Early Papermaking With Emphasis on Europe and the Fifteenth Century.* Kent, Ohio: Aspects of Librarianship No. 3, Kent State University.

117 J.W. Butler Paper Company. 1931. *The Story of Papermaking: an Account of Papermaking From Its Earliest Known Record Down to the Present Time.* Chicago, Illinois: J.W. Butler Paper Company.

118 Kay, John. 1893. *Paper, Its History.* London, England: Smith, Kay & Company.

119 Kazmeier, August Wilhelm. 1953. "Ein Druck auf Asbestpapier Über Asbestpapier." *Gutenberg-Jahrbuch* 28: 25-29.

120 _____. 1951. "Kork Als Ersatz für Papier." *Papiergeschichte* 1 (October): 45-46.

121 Keim, Betriebsleiter K. 1948. "Geschichtliche und Technische Entwicklung der Papiermacherei." *Wochenblatt für Papierfabrikation* 76 (December): 299-302.

122 Kellogg, Royal S. 1925. "Paper Making, From Papyrus to the Printed Page." *Mentor* 13 (September): 18-26.

123 Kent, Norman. 1967. "A Brief History of Papermaking." *American Artist* 31 (October): 36-41, 82.

124 Kenyon, Frederic G. 1939. *Papyrus: Alte Bücher und Moderne Entdeckungen.* Brunn, Germany: Rudolf M. Rohrer-Verlag.

125   Kersten, Paul. 1939. "Einband und Papier." *Philobiblon* 11 (March): 93-94.

126   ————. 1938. "Die Geschichte des Buntpapieres." *Wochenblatt für Papierfabrikation* 69 (November 12): 953-956.

127   ————. 1938. "Die Geschichte des Buntpapieres." *Wochenblatt für Papierfabrikation* 69 (November 19): 976-979.

128   ————. 1939. "Die Leimung Bei der Papierherstellung." *Philobiblon* 11 (March): 96.

129   ————. 1938. "Marmoriertes Buntpapier." *Wochenblatt, für Papierfabrikation* 69 (April 2): 299-300.

130   Kirschner, E. 1919. "Geschichte. Wert der Altpapiersendungen." *Wochenblatt für Papierfabrikation* 50 (April 19): 918.

131   Kohlmann, Erwin. 1968. "Die Kartenmacherei und der Inkunabelholzschnitt." *Marginalien* No. 29 (April): 15-29.

132   Kotte, Hans. 1956. "A History of Vegetable Parchment." *The Paper Maker* 25 (February): 9-14.

133   Kunze, F. 1913. "Das Papier im 17. Jahrhundert." *Der Papier-Fabrikant* 11 (March 28): 357-360.

134   Labarre, E.J. 1949. "The Sizes of Paper, Their Names, Origin & History." In *Buch and Papier,* edited by Horst Kunze. Leipzig, Germany: Otto Harrossowitz. 35-54.

135   Latour, A. 1949. "Paper, a Historical Outline." *Ciba Review* 6 (February): 2630-2639.

136   ————. 1949. "Uit de Geschiedenis van het Papier (I)." *De Papierwereld* 3 (January): 278-281.

137   ————. 1949. "Uit de Geschiedenis van het Papier (Slot)." *De Papierwereld* 3 (February): 298-300.

138   Le Clert, Louis. 1926. *Le Papier; Recherches et Notes Pour Servir à l'Histoire du Papier.* Paris, France: A l'Enseigne du Pégase. 2 Volumes.

139   Library of Congress. 1968. *Papermaking: Art and Craft.* Washington, D.C.: Library of Congress.

140   Lieuwes, Lieuwe. 1961. "Normalisierte Papierformate—Wahrhert und Dichtung." *Papiergeschichte* 11 (December): 74-77.

141   Loeber, E.G. 1972. "Tapa Jako Poprzednik Papieru." *Przegląd Papierniczy* 28 (November): 395-399.

142   Macdonald, R.G. 1953. "Paper: 25 Million Tons of Words a Year." *Print* 8 (June): 29-34.

143   Maddox, Harry A. 1939. *Paper; Its History, Sources, and Manufacture.* London, England: Sir I. Pitman & Sons. 6th Edition.

144 Mason, John. 1959. *Paper Making As an Artistic Craft.* London, England: Faber and Faber.

145 Morris, Henry. 1971. *The Bird & Bull Commonplace Book.* North Hills, Pennsylvania: Bird & Bull Press.

146 Morris, Henry, editor. 1963. *Five on Paper; a Collection of Five Essays on Papermaking, Books and Relevant Matters.* North Hills, Pennsylvania: Bird & Bull Press.

147 Morris, Henry. 1974. *The Paper Maker: a Survey of Lesser-Known Hand Paper Mills in Europe and North America.* North Hills, Pennsylvania: Bird & Bull Press.

148 Munsell, Joel. 1876. *Chronology of the Origins and Progress of Paper and Papermaking.* Albany, New York: J. Munsell. 5th Edition.

149 Opitz, A. 1974. "Früheste Beispiele für Papier Als Beschreibstoff." *Papiergeschichte* 24 (November): 23-24.

150 Piccard, Gerhard. 1961. "Das Alter der Spielkarten und die Papierforschung. Bedenken Gegen H. Rosenfelds These." *Archiv für Geschichte des Buchwesens* 3: 555-561.

151 _____. 1969. "II. Nochmals: Mühle und Blöwi (Order Plauel)." *Papiergeschichte* 19 (June): 11-15.

152 _____. 1965. "Die Papiergeschichtliche Sammlung und Ihre Ordnung." *Papiergeschichte* 15 (April): 14-21.

153 _____. 1968. "Riesaufdrucke und Riesumschläge." *Papiergeschichte* 18 (April): 1-26.

154 Pleziowa, Janina. 1949. "Dawni Antorzy o Papiernictwie." *Przegląd Papierniczy* 5 (July-August): 133-134.

155 Pollard, Graham. 1942. "Notes on the Size of the Sheet." *The Library (Fourth Series)* 22: 106-137.

156 Raithelhuber, Ernst. 1965. "Die Ersten Rundsiebpapier - und Pappenmaschinen der Geschichte." *Papiergeschichte* 15 (December): 59-73.

157 Rambaudi, P. 1947. "Observations on XV Century Printing Papers." *Paper & Print* 20 (Spring): 18, 21-22, 24, 26, 29.

158 Renker, Armin. 1954. "Bett der Letter. Reflexionen Über das Papier." *Druckspiegel* 9: 245-249.

159 _____. 1951. *Das Buch vom Papier.* Leipzig, Germany: Insel-Verlag zu Leipzig. 4th Edition.

160 _____. 1926. "Büttenpapier Als Druckpapier." *Archiv für Buchgewerbe und Gebrauchsgraphik* 63: 286-291.

161 _____. 1952-1953. " 'Byblis'." Ein Literarischer Beitrag zur

Andwendung der Beschreibstoffe vor dem Gebrauch des Papiers." *Imprimatur* 11: 188-190.

162 ————. 1956. "Eine Papierfabrik vor Hundert Jahren." *Gutenberg-Jahrbuch* 31: 57-61.

163 ————. 1935. "Der Papierliebhaber." *Philobiblon* 8 (January): 15-22.

164 ————. 1934. "Papierliebhaber in Allen Zeiten und Ständen." *Sankt Wiborada* 2: 68-79.

165 ————. 1925. "Papiermacher und Drucker." *Buch und Werbekunst* 2: 224-227.

166 ————. 1934. *Papiermacher und Drucker: ein Gespräch Über Alte und Neue Dinge.* Mainz, Germany: Drucker der Mainzer Presse.

167 ————. 1925-1926. "Spruch und Ausspruch Über Papier und Buch." *Monatsblätter für Bucheinbände und Handbindekunst* 2: 3-9.

168 Rosenfeld, Hellmut. 1961. "Die Spielkarten Als Volkstümliche Massenkunst." *Archiv für Geschichte des Buchwesens* 3: 561-566.

169 Santifaller, Leo. 1964. "Über Papierrollen Als Beschreibstoff." *Papiergeschichte* 14 (December): 49-56.

170 Sayce, A.H. 1872. "The Use of Papyrus As a Writing Material Among the Accadians." *Transactions of the Society of Biblical Archaeology* 1: 343-345.

171 Schäfer, G. 1949. "The Development of Papermaking." *Ciba Review* 6 (February): 2641-2648.

172 ————. 1949. "De Ontwikkeling der Papierfabricage (I)." *De Papierwereld* 3 (February): 318-320.

173 ————. 1949. "De Ontwikkeling der Papierfabricage (II)." *De Papierwereld* 3 (March): 338-340.

174 ————, and A. Latour. 1949. "The Paper Trade Before the Invention of the Paper-Machine." *Ciba Review* 6 (February): 2650-2656.

175 ————, and ————. 1949. "De Papiernijverheid Voor de Uitvinding van de Papiermachine." *De Papierwereld* 3 (April): 380-381.

176 ————, and ————. 1949. "De Papiernijverheid Voor de Uitvinding van de Papiermachine (Slot)." *De Papierwereld* 3 (April): 398-400.

177 Schlieder, Wolfgang. 1960. "Einige Bemerkungen Über die

Entwicklung des Papierbedarfs." *Papiergeschichte* 10 (December): 80-84.

178 ————. 1959. "Aus der Jugend der Papiermaschine." *Zellstoff und Papier* 8 (February): 61-63.

179 Schmidt-Kunsemüller, F.A. 1957. "... Und Schöpft ein Blatt, einen Bogen ein Lied vom Papier." *Papiergeschichte* 7 (April): 26-28.

180 Schulte, Alfred. 1938. "Hundert Jahre Patentpapierfabrik Hohenofen." *Der Papier-Fabrikant* 36 (August 12): 363-364.

181 ————. 1938. "Der Papiermacher und die Papiergeschichte." *Der Papier-Fabrikant* 36 (February 18): 68-72.

182 ————. 1938. "Papiermacherei Seit vor 1600." *Der Altenburger Papier* 12: 985-986.

183 ————. 1939. "Papierpresse, Druckerpresse und Kelter." *Gutenberg-Jahrbuch* 14: 52-56.

184 ————. 1933. "Der Ursprüngliche Name Unseres Holländers." *Der Papier-Fabrikant* 31 (May 21): 305-306.

185 ————, edited by Toni Schulte. 1955. *Wir Machen die Sachen die Nimmer Vergehen.* Wiesbaden, Germany: Das Betriebliche Leben Industrie-Verlags-GMBH Dr. Edgar Jörg.

186 Schulte, Toni. 1953. "Ein Umstrittenes Wappen aus dem 16. Jahrhundert." *Papiergeschichte* 3 (November): 59-62.

187 Schulze, Bruno. 1928. "Beitrag zur Ältesten Geschichte der Papiererzeugung." *Wochenblatt für Papierfabrikation* 59 (March 10): 257-260.

188 Sembritzki, Walter. 1931. "Ueber die Erste Papiermaschine und Ihre Erfindung." *Der Papier-Fabrikant* 29 (September 27): 625-627.

189 Smith, Mason Rossiter. 1974. "Papyrus." *Scholarly Publishing* 6 (October): 86-89.

190 Šorn, Jože. 1956. "Ältere Papiermühlen in Slowenien." *Papiergeschichte* 6 (July): 40-42.

191 Spicer, A.D. 1907. *The Paper Trade.* London, England: Methuen.

192 Sporhan-Krempel, Lore. 1955. "Pergamenter und Papier." *Wochenblatt für Papierfabrikation* 83 (December 15): 981-982.

193 Szönyi, J. László. 1908. *14. Szazadbeli Papiros-Okleveleink Vizjegyei.* Budapest, Hungary: Athenaeum Irodalmi és Nyomdai Részvénytársulat.

194 Thiel, Viktor. 1936. "Zur Frühgeschichte der Papiererzeugung." *Wochenblatt für Papierfabrikation* 67 (Special Edition): 9-14.

195 _____. 1936. "Zur Frühgeschichte der Papiererzeugung."
Wochenblatt für Papierfabrikation 67 (December 5): 910-915.
196 _____. 1935. "Die Geschichtliche Sendung des Papiers."
Wochenblatt für Papierfabrikation 66 (Special Edition): 3-6.
197 _____. 1950. "Riesumschläge aus dem Oberen Donauraum."
Gutenberg-Jahrbuch 25: 46-50.
198 _____. 1932. "Die Rolle des Papiers in der Kulturellen
Entwicklung der Menschheit." *Wochenblatt für
Papierfabrikation* 63 (June 4A): 10-13.
199 Thompson, Lawrence. 1938. *The Development of the Book:
I. Writing Materials 3500 B.C.-A.D. 1500.* Princeton, New
Jersey: Bulletin No. 1, Princeton University Library.
200 Tomaszewska, Wanda. 1968. "O Papierze, Który Zyskał Miano
Wiecznotrwałego." *Przegląd Papierniczy* 24 (August): 290-291.
201 Tommasini, A.R. 1963. *The Story of Paper, Told Briefly Once
Again.* Millbrae, California: Privately Printed by A.R.
Tommasini.
202 Treumann, Hans. 1940. "Ein Riesumschlag aus Räpina." *Gutenberg-
Jahrbuch* 15: 51-53.
203 Tschudin, Peter. 1965. "Papyrus." *The Paper Maker* 34 (March):
3-15.
204 _____. 1952. "Woher Stammt das Wort 'Papier'."
Papiergeschichte 2 (December): 79-82.
205 Tschudin, W. Fritz. 1964. "Von Alten Papiermacherpflanzen."
Textil-Rundschau 19 (December): 648-655.
206 _____. 1965. "Von Alten Papiermacherpflanzen." *Textil-
Rundschau* 20 (October): 309-318.
207 Volkmann, Kurt. 1935. "Beiträge zur Geschichte der
Papierliebhaberei." *Zeitschrift für Büchfreunde (Third Series)*
4 (January): 5-9.
208 Voorn, Henk. 1961. "A Brief History of the Sizing of Paper." *The
Paper Maker* 30 (February): 47-53.
209 _____. 1955. "Zur Erfindung des Holländers." *Papiergeschichte*
5 (July): 38-42.
210 _____. 1950. "De Geschiedenis van de Hollander." *De
Papierwereld* 4 (July): 437-439.
211 _____. 1952. "In Search of New Raw Materials: Being the Narra-
tion of the Many Efforts of Papermakers, Clergymen, and
Scholars To Make Paper From Materials Other Than Rags, and
of the Curious Books They Left Us." *The Paper Maker* 21
(September): 1-14.

212   _____ . 1968. "New Thoughts On Old Papermaking." *De Papierwereld* 23 (August): 216-219.

213   _____ . 1968. "New Thoughts On Old Papermaking." *De Papierwereld* 23 (September): 239-245.

214   _____ . 1969. *Old Ream Wrappers: an Essay on Early Ream Wrappers of Antiquarian Interest.* North Hills, Pennsylvania: Bird & Bull Press.

215   _____ . 1956. "On the Invention of the Hollander Beater." *The Paper Maker* 25 (September): 1-9.

216   _____ . 1950. "Over Het Satineren van Papier." *De Papierwereld* 4 (May): 293-295.

217   _____ . 1968. "Paper: Instrument of Liberty, Pacemaker of Progress, Support of Civilization." *The Quarterly Journal of the Library of Congress* 25 (April): 105-115.

218   _____ . 1950. "De Papiermaker." *De Papierwereld* 5 (November): 148-152.

219   _____ . 1949. "Senger's 'Älteste Urkunde der Papierfabrikation'." *De Papierwereld* 4 (October): 126-128.

220   _____ . 1958. "A Short History of the Glazing of Paper." *The Paper Maker* 27 (February): 3-10.

221   _____ . 1952. "Über Léorier de l'Isle." *Papiergeschichte* 2 (December): 75-79.

222   _____ . 1953. "Wann Wurde der Holländer Erfunden?" *Papiergeschichte* 3 (December): 84.

223   Weaver, Alexander. 1937. *Paper, Wasps, and Packages: the Romantic Story of Paper and Its Influence on the Course of History.* Chicago, Illinois: Container Corporation of America.

224   Weirich, Hans. 1931. "Was Ist Büttenpapier?" *Zeitschrift für Buchfreunde (New Series)* 23 (April-May): 95-96.

225   Weiss, Karl Theodor. 1925. "Vom Kupferdruckpapier." *Wochenblatt für Papierfabrikation* 56 (November 7): 1366-1368.

226   _____ . 1925. "Vom Kupferdruckpapier." *Wochenblatt für Papierfabrikation* 56 (December 19): 1536-1539.

227   _____ . 1920. "Das Papier in Spruch und Sprache." *Wochenblatt für Papierfabrikation* 51 (March 27): 863.

228   _____ . 1920. "Das Papier in Spruch und Sprache." *Wochenblatt für Papierfabrikation* 51 (June 5): 1563.

229   _____ . 1920. "Das Papier in Spruch und Sprache." *Wochenblatt für Papierfabrikation* 51 (July 10): 1919.

230   _____ . 1920. "Das Papier in Spruch und Sprache." *Wochenblatt für Papierfabrikation* 51 (July 17): 1999.

231 ————. 1920. "Das Papier in Spruch und Sprache." *Wochenblatt für Papierfabrikation* 51 (August 7): 2190.

232 ————. 1920. "Das Papier in Spruch und Sprache." *Wochenblatt für Papierfabrikation* 51 (September 4): 2462.

233 ————. 1920. "Das Papier in Spruch und Sprache." *Wochenblatt für Papierfabrikation* 51 (October 2): 2753.

234 ————. 1920. "Das Papier in Spruch und Sprache." *Wochenblatt für Papierfabrikation* 51 (November 6): 3108.

235 ————. 1921. "Das Papier in Spruch und Sprache." *Wochenblatt für Papierfabrikation* 52 (January 8): 27.

236 ————. 1921. "Das Papier in Spruch und Sprache." *Wochenblatt für Papierfabrikation* 52 (February 5): 339.

237 ————. 1921. "Das Papier in Spruch und Sprache." *Wochenblatt für Papierfabrikation* 52 (February 19): 495.

238 ————. 1921. "Das Papier in Spruch und Sprache." *Wochenblatt für Papierfabrikation* 52 (March 26): 910.

239 ————. 1921. "Das Papier in Spruch und Sprache." *Wochenblatt für Papierfabrikation* 52 (March 31): 991.

240 ————. 1921. "Das Papier in Spruch und Sprache." *Wochenblatt für Papierfabrikation* 52 (April 16): 1149.

241 ————. 1921. "Das Papier in Spruch und Sprache." *Wochenblatt für Papierfabrikation* 52 (August 27): 2766.

242 ————. 1921. "Das Papier in Spruch und Sprache." *Wochenblatt für Papierfabrikation* 52 (September 24): 3104.

243 ————. 1921. "Das Papier in Spruch und Sprache." *Wochenblatt für Papierfabrikation* 52 (September 30): 3174.

244 ————. 1921. "Das Papier in Spruch und Sprache." *Wochenblatt für Papierfabrikation* 52 (October 15): 3356.

245 ————. 1921. "Das Papier in Spruch und Sprache." *Wochenblatt für Papierfabrikation* 52 (November 12): 3686.

246 ————. 1921. "Das Papier in Spruch und Sprache." *Wochenblatt für Papierfabrikation* 52 (November 19): 3791.

247 ————. 1921. "Das Papier in Spruch und Sprache." *Wochenblatt für Papierfabrikation* 52 (November 26): 3882.

248 ————. 1921. "Das Papier in Spruch und Sprache." *Wochenblatt für Papierfabrikation* 52 (December 17): 4161.

249 ————. 1921. "Das Papier in Spruch und Sprache." *Wochenblatt für Papierfabrikation* 52 (December 24): 4250.

250 ————. 1922. "Das Papier in Spruch und Sprache." *Wochenblatt für Papierfabrikation* 53 (January 7): 32.

251 _____. 1922. "Das Papier in Spruch und Sprache." *Wochenblatt für Papierfabrikation* 53 (February 4): 406.

252 _____. 1922. "Das Papier in Spruch und Sprache." *Wochenblatt für Papierfabrikation* 53 (March 4): 752.

253 _____. 1922. "Das Papier in Spruch und Sprache." *Wochenblatt für Papierfabrikation* 53 (March 11): 853.

254 _____. 1922. "Das Papier in Spruch und Sprache." *Wochenblatt für Papierfabrikation* 53 (March 25): 1052.

255 _____. 1922. "Das Papier in Spruch und Sprache." *Wochenblatt für Papierfabrikation* 53 (April 15): 1338.

256 _____. 1922. "Das Papier in Spruch und Sprache." *Wochenblatt für Papierfabrikation* 53 (May 13): 1720.

257 _____. 1922. "Das Papier in Spruch und Sprache." *Wochenblatt für Papierfabrikation* 53 (June 17): 2200.

258 _____. 1925. "Das Papier in Spruch und Sprache." *Wochenblatt für Papierfabrikation* 56 (December 26): 1596.

259 _____. 1926. "Das Papier in Spruch und Sprache." *Wochenblatt für Papierfabrikation* 57 (February 13): 184.

260 _____. 1926. "Das Papier in Spruch und Sprache." *Wochenblatt für Papierfabrikation* 57 (February 27): 253.

261 _____. 1926. "Das Papier in Spruch und Sprache." *Wochenblatt für Papierfabrikation* 57 (May 15): 576.

262 _____. 1927. "Das Papier in Spruch und Sprache." *Wochenblatt für Papierfabrikation* 58 (February 5): 134.

263 _____. 1928. "Das Papier in Spruch und Sprache." *Wochenblatt für Papierfabrikation* 59 (January 14): 50.

264 _____. 1928. "Das Papier in Spruch und Sprache." *Wochenblatt für Papierfabrikation* 59 (February 11): 147.

265 _____. 1928. "Das Papier in Spruch und Sprache." *Wochenblatt für Papierfabrikation* 59 (September 8): 1010.

266 _____. 1929. "Das Papier in Spruch und Sprache." *Wochenblatt für Papierfabrikation* 60 (June 22): 782.

267 _____. 1930. "Das Papier in Spruch und Sprache." *Wochenblatt für Papierfabrikation* 61 (May 10): 598.

268 _____. 1930. "Das Papier in Spruch und Sprache." *Wochenblatt für Papierfabrikation* 61 (June 28): 848.

269 _____. 1930. "Das Papier in Spruch und Sprache." *Wochenblatt für Papierfabrikation* 61 (July 5): 880.

270 _____. 1930. "Das Papier in Spruch und Sprache." *Wochenblatt für Papierfabrikation* 61 (August 2): 1017.

271 ————. 1931. "Das Papier in Spruch und Sprache." *Wochenblatt für Papierfabrikation* 62 (May 9): 675.
272 ————. 1931. "Das Papier in Spruch und Sprache." *Wochenblatt für Papierfabrikation* 62 (August 8): 773.
273 ————. 1931. "Das Papier in Spruch und Sprache." *Wochenblatt für Papierfabrikation* 62 (August 15): 796.
274 ————. 1931. "Das Papier in Spruch und Sprache." *Wochenblatt für Papierfabrikation* 62 (August 29): 841.
275 ————. 1931. "Das Papier in Spruch und Sprache." *Wochenblatt für Papierfabrikation* 62 (December 19): 1208.
276 ————. 1932. "Das Papier in Spruch und Sprache." *Wochenblatt für Papierfabrikation* 63 (May 7): 363.
277 ————. 1932. "Das Papier in Spruch und Sprache." *Wochenblatt für Papierfabrikation* 63 (June 11): 474.
278 ————. 1932. "Das Papier in Spruch und Sprache." *Wochenblatt für Papierfabrikation* 63 (October 29): 822.
279 ————. 1949. "Vom Velin- und Kupferdruckpapier." *Wochenblatt für Papierfabrikation* 77 (December): 483-486.
280 ————. 1942-1943. "Zeilenpapier." *Buch und Schrift* 5-6: 106-158.
281 Weiss, Wisso. 1965. "Buchdrucker Erhatten die Kontrolle Über das Lumpensammeln." *Gutenberg-Jahrbuch* 40: 13-17.
282 ————. 1955. "Vom Doppelpapier." *Gutenberg-Jahrbuch* 30: 19-21.
283 ————. 1962. "Zur Geschichte des Löschpapiers." *Gutenberg-Jahrbuch* 37: 13-18.
284 ————. 1949. "Der Postreiter Mit der Friedensbotschaft." *Allgemeine Papier-Rundschau* No. 4 (June 15): 166-167.
285 ————. 1963. "Vom Titelpapier." *Gutenberg-Jahrbuch* 38: 11-16.
286 ————. 1956. "Das Wanderbuch Eines Papiermachergesellen aus Zittau." *Papiergeschichte* 6 (December): 76-84.
287 ————. 1951. "Zierrand-Papier." *Gutenberg-Jahrbuch* 26: 40-47.
288 Wheelock, Mary E. 1928. *Paper: Its History and Development.* Chicago, Illinois: American Library Association.
289 Wheelwright, W.B. 1939. "Pulp and Paper: Its Past, Present & Future." *The Paper Maker* 8 (February): 10-12.
290 ————. 1939. "Pulp and Paper: Its Past, Present & Future." *The Paper Maker* 8 (June): 11-13.
291 Wiesner, Jules. 1887. "Die Faijûmer und Uschmûneiner Papiere."

Mittheilungen aus der Sammlung der Papyrus Erzherzog Rainer
2-3: 179-260.

292 _____. 1887. "Mikroskopische Untersuchung der Papiere von el
Faijûm." *Mittheilungen aus der Sammlung der Papyrus
Erzherzog Rainer* 1: 45-48.

293 _____. 1903-1904. "Ein Neuer Beitrag zur Geschichte des
Papiers." *Sitzungsberichte der Kaiserlichen Akademie der
Wissenschaften* 148:5.

294 Wilisch, Gottfried Julius Alexander. 1938. "Die Geschichte des
Buntpapieres." *Wochenblatt für Papierfabrikation* 69
(December 10): 1086-1087.

295 Williamson, K. 1944. "Paper Making in Man; the Story of a Lost
Industry." *World's Paper Trade Review* 121 (June 9): 1307-
1308, 1310, 1342, 1344.

296 Zuman, F. 1938. "Die Anfänge und die Entwicklung der
Papiermacherei in Europa." *Der Altenburger Papierer* 12:
298-301, 382-385, 460-463, 536-546.

297 _____. 1934. "Z Dějin Papírnictví." *Český Časopis Historický*
40: 675-678.

298 _____. 1933. "Papírnictví Evropské." *Český Časopis Historický*
39: 678-680.

299 _____. 1931. "Počátky a Vývoj Papírnictví v Evropě." *Slovenská
Vlastivěda* 1: 1-20.

WATERMARKS

300 Abrams, T.M. 1963. "The History and Artistry of Watermarks."
Pulp and Paper Magazine of Canada 64 (November): 79-81.

301 Aiken, P.H. 1890. "Preliminary Notes on 15th Century Watermarks." *Transactions of the Glasgow Archaeological Society
(New Series)* 1: 535-539.

302 Alberti, K. 1927. "Die Ehemaligen Papiermühlen im Aschergebiete
und Ihre Papier-Wasserzeichen." *Unser England* 31: 69-77.

303 Antz, E.L. 1922. "Wasserzeichen und Fabrikmarken." *Der Papier-
Fabrikant* 20 (May 14): 619-623.

304 Bayley, Harold. 1912. *The Lost Language of Symbolism.* London,
England: Williams and Norgate.

305 _____. 1906. "Notes on Watermarks." *Booklovers' Magazine* 6:
65-71.

306 _____. 1909. "Papermarks." In *A New Light on the Renaissance*

Displayed in Contemporary Emblems, by H. Bayley. London, England: J. Dent & Company. 1-110.

307 _____. 1908. "The Romance of Watermarks." *The Bibliophile* 1 (April): 93-96.

308 Beadle, Clayton. 1906. "The Development of Watermarking in Hand- and Machine-Made Papers." *Journal of the Royal Society of Arts* 54 (May 18): 684-700.

309 _____. 1911. "The Development of Watermarking in Paper." *Paper* 3 (April 5): 9-12, 28, 30, 32.

310 _____. 1908. "The Study of Ancient Watermarks." *World's Paper Trade Review* 29 (April 3): 569-573.

311 Blades, William. 1889. "On Paper and Paper-Marks." *The Library (First Series)* 1: 217-223.

312 Blansch, J. le. 1953. "De Watermerken." *Vlaamse Drukker* 13 (February): 46-47.

313 Bofauill Y Sans, Francisco de. 1959. *Animals in Watermarks.* Hilversum, Holland: The Paper Publications Society.

314 Bogdán, István. 1956. "Vizjelek és Vizjelkutatás." *Livéltári Hiradó* 8: 27-35.

315 _____. 1956. "Vizjelek és Vizjelkutatás." *Papír-és Nyomdatechnika* 8 (January): 25-28.

316 _____. 1974. "Vízjellel az Iralhamisítás Ellen." *Papíripar* 18: 30-36.

317 Bradley, W.A. 1910. "Early Watermarks." *Printing Art* 16 (November): 176-180.

318 Briquet, Charles Moïse. 1955. *Briquet's Opuscula: the Complete Works of Dr. C.M. Briquet.* Hilversum, Holland: The Paper Publications Society.

319 _____. 1900. "La Date de Trois Impressions Précisée Par Leurs Filigranes." *Le Bibliographe Moderne* 4: 113-124.

320 _____. 1892. "De la Valeur des Filigranes de Papier Comme Moyen de Déterminer l'Âge et la Provenance des Documents Non Datés." *Bulletin de la Société d'Histoire et d'Archéologie de Genève* 1: 192-202.

321 Bühler, Curt F. 1973. "Last Words on Watermarks." *The Papers of the Bibliographical Society of America* 67 (First Quarter): 1-16.

322 _____. 1957. "Watermarks and the Dates of Fifteenth-Century Books." *Studies in Bibliography* 9: 217-224.

323 Bureau, W. 1958. "Paper: Watermarks." *Graphic Arts Monthly* 30 (August): 42, 44.

324 Clapperton, R.H. 1950. "The First Dandy Roll." *World's Paper Trade Review* 134 (August 10): 402,404.

325 Connelly, F. 1947. "Historical Aspects of Watermarking." *Paper Making* 66 (Autumn): 50.

326 Elliott, Harrison. 1948. "The Romantic Phase of Watermarks in Paper With Something of Their Significance as Symbols." *The Paper Maker* 17 (September): 11-15.

327 Fiskaa, H.M. 1939. "Vannmerkeforskningen og dens Betydning For Bestemmelsen av Gamle Bøker or Håndskrifter." *Nordisk Tidskrift för Bok- och Biblioteksväsen* 26 (October-December): 201-223.

328 Flockhart, T.A. 1942. "Watermarks." *Paper & Paper Products* 83 (January 20): 6.

329 Gachet, Henri. 1952. "The Mystery and Beauty of Watermarks." *Papetier* 6 (December): 53, 55.

330 _____. 1954. "Des Premiers Papiers aux Premiers Filigranes." *Le Courrier Graphique* No. 72 (May): 27-36.

331 Gasparinetti, A.F. 1958. "Eine Bestellung von Wasserzeichenpapier in Alter Zeit." *Papiergeschichte* 8 (July): 40-43.

332 _____. 1958. "Briquet Unbekannt Gebliebene Frühe Autoren, die Über Wasserzeichen Geschrieben Haben." *Papiergeschichte* 8 (December): 71-74.

333 _____. 1964. "A Curiosity of Papermaking History." *The Paper Maker* 33 (September): 33-38.

334 _____. 1952. "Über die 'Entstellung' (Bedentungs-und Formwandel) von Wasserzeichen." *Papiergeschichte* 2 (June): 33-36.

335 Gerardy, Theodor. 1965. "Der Aufbau Einer Wasserzeichensammlung." *Papiergeschichte* 15 (April): 7-14.

336 _____. 1964. *Datieren Mit Hilfe von Wasserzeichen; Beispielhaft Dargestelt an der Gesemtproduktion der Schaumburgischen Papiermühle Ahrensburg von 1604-1650.* Bückeburg, Germany: Verlag Grimme Bückeburg.

337 Grosse-Stoltenberg, Robert. 1961. "Wasserzeichen in Alten Landkarten." *Papiergeschichte* 11 (December): 93-96.

338 Grozdanović-Pajić, M. 1968. "Vodeni Znak 'Tri Polumeseca'." *Bibliotekar* 5: 527-550.

339 Heawood, Edward. 1950. *Historical Review of Watermarks.* Amsterdam, The Netherlands: Swets and Zeitlinger.

340 _____. 1928. "The Position on the Sheet of Early Watermarks." *The Library (Fourth Series)* 9 (June): 38-47.

341 _____. 1924. "The Use of Watermarks in Dating Old Maps and Documents." *The Geographical Journal* 63 (May): 391-412.

342 _____. 1950. *Watermarks*. Hilversum, Holland: The Paper Publications Society.

343 Hegg, Peter. 1954. "Ein Unikum?" *Schweizerisches Gutenbergmuseum* 40: 3-8.

344 Henderson Aitken, P. 1890. "A Preliminary Note on 15th. Century Watermarks." *Transactions of the Glasgow Archaeological Society* 1.

345 Hendrich, W.G. 1952. "Watermarking." *Pulp and Paper Magazine of Canada* 53 (June): 108-110.

346 Horodisch, A. 1952. "On the Aesthetics of Ancient Watermarks." In *The Briquet Album,* edited by E.J. Labarre. Hilversum, Holland: The Paper Publications Society. 107-117.

347 Hössle, F. von. 1914. "Karneval im Verborgenen Oder der Harlekin im Papier." *Der Papier-Fabrikant* 12 (June 19A): 34-41.

348 Hunter, Dard. 1921. "Old Watermarks of Animals." *Paper* 28 (August 24): 12-15, 25.

349 _____. 1922. "Portrait Watermarks." *Paper* 29 (January 11): 13-14.

350 _____. 1938. "Romance of Watermarks." *Paper Progress* 1 (April): 9-12, 36-38.

351 _____. 1939. *Romance of Watermarks: a Discourse on the Origin and Motive of These Mystic Symbols Which First Appeared in Italy Near the End of the 13th Century.* Cincinnati, Ohio: The Stratford Press.

352 _____. 1923. "Symbolism in Paper Markings." *Paper* 30 (December 20): 3-6.

353 _____. 1936. "The Use and Significance of the Ancient Watermarks." *Paper & Printing Digest* 2 (December): 3-7.

354 _____. 1921. "Watermarking Handmade Papers." *Scientific American* 124 (March 26): 248-249.

355 _____. 1940. "The Watermarking of Portraits, Ancient and Modern." *Indian Print and Paper* 5 (March): 22-25.

356 Irigoin, Jean. 1966. "Groupes et Séries de Filigranes au Début du XIVe Siècle." *Papiergeschichte* 16 (December): 18-22.

357 Keinz, Friedrich. 1897. "Über die Älteren Wasserzeichen des Papiers und Ihre Untersuchung." *Zeitschrift für Bücherfreunde* 1 (August): 240-247.

358 Klepikov, S.A. 1958. "Bumaga s Filigrań yu 'Gerb Goroda

Amsterdama'." *Zapiski Otdela Rukopiseǐ Gos. Biblioteki SSSR im V.I. Lenina* 20: 315-352.

359 Korn, R. 1950. "Normal-Wasserzeichenpapiere." *Allgemeine Papier-Rundschau* No. 3: 94-95.

360 Krisch, William. 1903. "The Raison d'Être of Mediaeval Paper-marks." *Baconiana (Third Series)* 1: 225-235.

361 Latour, A. 1949. "Het Watermerk." *De Papierwereld* 3 (May): 442-443.

362 Liljedahl, Gösta. 1958. "Nyare Litteratur Rörande Vattenmärken." *Historisk Tidskrift (Second Series)* 21 (April-June): 206-220.

363 _____. 1970. "Om Vattenmärken i Papper och Vattenmärks-forskning (Filigranologi)." *Biblis* n.v.: 91-129.

364 _____. 1956. "Om Vattenmärken och Filigranologi." *Historisk Tidskrift (Second Series)* 19 (July-September): 241-274.

365 _____. 1956. "Wasserzeichen Als Beweismittel vor Gericht." *Papiergeschichte* 6 (December): 75-76.

366 Mackenzie, J.G. 1945. "Watermarks, a Brief History of Some Early Designs." *Paper & Print* 18 (Winter): 192-194, 196.

367 Martin, E. 1961. "Das Wasserzeichen im Papier Als Kriminalistisches Spurenelement." *Internationale Kriminalistische Revue* 16: 205-211.

368 Martin, H.R. 1957. "Watermarks." *Paper Making* 76 (Summer): 6-16.

369 Meldau, Robert. 1940. "Zur Bedeutung der Hand Alswasserzeichen." *Gutenberg-Jahrbuch* 15: 41-50.

370 Mošin, Vladimir. 1973. *Anchor Watermarks.* Amsterdam, The Netherlands: The Paper Publications Society.

371 _____, and Mira Grozdanović-Pajić. 1967. *Agneau Pascal.* Belgrade, Yugoslavia: Editions 'Prosveta'.

372 _____, and Seid Traljić. 1957. *Filigranes des XIIIe et XIVe SS.* Zagreb, Yugoslavia: Académie Yougoslave das Sciences et des Beauz-Arts, Institute d'Histoire. 2 Volumes.

373 Nordstrand, Ove K. 1970. "Vandmaerker og Vandmaerkeforskning. Papirhistoriske Noter, I." *Fund og Forskning* 17: 7-20.

374 Piccard, Gerhard. 1960. "Die Datierung des Missale Speciale (Constantiense) Durch Seine Papiermarken." *Archiv für Geschichte des Buchwesens* 2: 571-584.

375 _____. 1956. "Die Wasserzeichenforschung Als Historische Hilfswissenschaft." *Archivalische Zeitschrift* 52: 62-115.

376 _____. 1955. "Wasserzeichenkunde und Urbarforschung." *Archivum* 2: 65-81.

377   Platbarzdis, A. 1955. "Das Erste Wasserzeichen zum Kennzeichnen von Banknotenpapier." *Papiergeschichte* 5 (November): 62-66.

378   Redgrave, Gilbert R. 1904. "The Water-Marks in Paper." *The Library (Second Series)* 5 (January): 91-92.

379   Renker, Armin. 1952. "Art of the Watermark." *Graphis* 8 (No. 39): 52-61, 92.

380   _____. 1950. "Feinpaper und Wasserzeichen." *Allgemeine Papier-Rundschau* No. 3 (February 15): 95-96.

381   _____. 1957. "Filigranophilie." *Philobiblon (Hamburg)* 1 (September): 231-234.

382   _____. 1937. "Sinn und Bedeutung der Wasserzeichen." *Der Papier-Fabrikant* 35 (August 6): 305-309.

383   _____. 1952. "Sinn und Bedeutung der Wasserzeichen im Papier Einst und Jetzt." *Der Büromarkt* 7: 610-614.

384   _____. 1936-1937. "Das Wasserzeichen Als Kulturspiegel." *Imprimatur* 7: 176-190.

385   _____. 1927. "Das Wasserzeichen ein Entlegenes Feld Bibliophiler Betätigung." *Zeitschrift für Büchfreunde (New Series)* 19 (April): 61-66.

386   _____. 1925. "Wasserzeichen in Alter und Neuer Zeit." *Buch und Werbekunst* 2: 549-553.

387   _____. 1930. "Wasserzeichen, Kulturgut des Papiermachers im Mittelalter." *Buch und Werbekunst* 7: 124-128.

388   Schulte, Alfred. 1938. "Dürerforschung und Wasserzeichen." *Der Papier-Fabrikant* 36 (March 25): 110.

389   _____. 1934. "Papiermühlen = und Wasserzeichenforschung." *Gutenberg-Jahrbuch* 9: 9-27.

390   _____. 1934. "Schöpfformen und Doppelformen." *Wochenblatt für Papierfabrikation* 65 (October 13): 724-726.

391   _____. 1939. "Wasserzeichenfunde aus dem Altpapier." *Der Papier-Fabrikant* 37 (July 28): 267.

392   Schulte, T. 1951. "Rätsel der Wasserzeichenforschung." *Papiergeschichte* 1 (May): 9-13.

393   Siegl, Karl. 1925. "Die Egerer Papierwasserzeichen." *Unser Egerland* 29: 3-7.

394   Smith, V.S. 1953. "An Introduction to a Subject of Absorbing Interest to Many Buyers of Printing; Watermarks." *British Printer* 66 (September-October): 55-59.

395   Stenger, Erich. 1929. "Objektive Feststellung von Wasserzeichen." *Der Papier-Fabrikant* 27 (June A): 84-87.

396   Stevenson, Allan. 1967. *The Problem of the Missale Speciale.*
        London, England: The Bibliographical Society.
397   _____. 1951-1952. "Watermarks Are Twins." *Studies in Bibli-
        ography* 4: 57-91.
398   Szönyi, J. László. 1908. *14. Szazadbeli Papiros-Okleveleink
        Vizjegyei.* Budapest, Hungary: Athenaeum Irodalmi és
        Nyomdai Reszvénytársulat.
399   Thayer, C.S. 1968. "Why Is a Watermark?" *Modern Lithography* 36
        (December): 43, 45, 48-49.
400   Thomas, Henry. 1938-1945. "Watermarks." *Edinburgh Biblio-
        graphical Society Transactions* 2: 449-450.
401   Vallet-Viriville, Auguste. 1859. "Notes Pour servir à l'Histoire du
        Papier." *Gazette des Beaux-Arts* 2: 222-236.
402   Voorn, Henk. 1950. "Het Fabeldier Als Watermerk: de Basilisk en
        de Draak." *De Papierwereld* 5 (October): 99-102.
403   _____. 1950. "Het Fabeldier Als Watermerk: de Eenhoorn."
        De Papierwereld 5 (September): 59-62.
404   _____. 1962. "Fabulous Beasts in Watermarks." *The Paper
        Maker* 31 (September): 1-9.
405   _____. 1957. "Fabulous Beasts in Watermarks: the Basilisk."
        The Paper Maker 26 (September): 19-24.
406   Weiss, Karl Theodor. 1957. "Beschreibung der Wasserzeichen."
        Zellstoff und Papier 6 (August): 255-256.
407   _____. 1953. "Die Bestimmung Stuttgarter Handschrift Mit
        Hilfe des Verwendeten Papieres und Seiner Wasserzeichen."
        Gutenberg-Jahrbuch 28: 16-24.
408   _____. 1962. *Handbuch der Wasserzeichenkunde.* Leipzig,
        Germany: VEB Fachbuchverlag Leipzig.
409   _____. 1926. "Papiergeschichte und Wasserzeichenkunde.
        Erreichte Ziele und Zu Lösende Aufgaben." *Archiv für
        Buchgewerbe* 63: 292-308.
410   Weiss, Wisso. 1958. "Eckzier-Wasserzeichen." *Gutenberg-Jahrbuch*
        33: 37-43.
411   _____. 1950. "Kulturgeschichtliche Betrachtung Zur
        Wasserzeichenfrage." *Allgemeine Papier-Rundschau* No. 4
        (February 28): 162-163.
412   _____. 1944-1949. "Das Posthorn: ein Beitrag Zur
        Wasserzeichenkunde." *Gutenberg-Jahrbuch* 19-24: 39-46.
413   _____. 1957. "Vom Stempelpapier und Seinem Wasserzeichen."
        Gutenberg-Jahrbuch 32: 26-32.

414 _____. 1967. "Über das Ordnen Einer Wasserzeichen- und Papiersammlung." *Papiergeschichte* 17 (June): 33-43.

415 _____. 1960. "Vom Wasserzeichen im Druckpapier." *Gutenberg-Jahrbuch* 35: 11-18.

416 _____. 1966. "Wasserzeichen im Maschinenpapier." *Jahrbuch der Deutschen Bücherei* 2: 93-111.

417 _____. 1961. "Das Wasserzeichen in Alten Handgeschöpften Velinpapier." *Gutenberg-Jahrbuch* 36: 11-17.

418 _____. 1961. "Watermarks in Old Handmade Wove Paper." *The Paper-Maker (London)* 142 (November): 62, 64-65.

419 Wheelwright, W.B. 1924. "The Origin of Watermarks." *Pulp and Paper Magazine of Canada* 22 (January 10): 42.

420 Zeitler, Julius. 1926. "Meister der Wasserzeichenkunde." *Der Papier-Fabrikant* 24 (November 21): 721-722.

2 *The History of*
Paper and Papermaking
in Asia
and Australia

421 Bockwitz, Hans H. 1950. "Dokuments Zur Papiergeschichte;
Quellen Zur Geschichte der Asiatischen Papiermacherei."
Wochenblatt für Papierfabrikation 78 (February 15): 67-68.

422 _____. 1939. "Die Drei Ältesten Papiermacher-'Handbücher'
des Orients." *Der Papier-Fabrikant* 37 (June 16): 207-212.

423 Hunter, Dard. 1940. "The Papermaking Moulds of Asia." *Gutenberg-
Jahrbuch* 15: 9-24.

424 _____. 1943. "Sacred Papers of the Orient." *Paper and Twine
Journal* 16 (February): 17-21.

AUSTRALIA

425 Loeber, E.G. 1958. "History of Paper-Making in Australia."
Papiergeschichte 8 (December): 74-77.

CEYLON

426 Nordstrand, Ove K. 1961. "The Introduction of Paper in Ceylon."
Papiergeschichte 11 (December): 67-70.

CHINA

427 Bockwitz, Hans H. 1942. "Eine Chinesische Papiermacher -
Werkstatt um 1800." *Archiv für Buchgewerbe und
Gebrauchsgraphik* 79: 451.

428 _____. 1941. "Fernöstliche Papiermacherei der Gegenwart."
Archiv für Buchgewerbe und Gebrauchsgraphik 78 (July):
246-252.

429 _____. 1939. "Hat die Orientalische Papiermacherei Bereits
Mechanische Stampfwerke Gekannt?" *Wochenblatt für
Papierfabrikation* 70 (April 22): 349-352.

430 _____. 1931. "Proben Ältesten Chinesischen Papiers aus dem
Zweiten Nachchristlichen Jahrhundert." *Archiv für
Buchgewerbe und Gebrauchsgraphik* 68: 707.

431 _____. 1942. "Die Rolle des Papiers bei Chinesischen
Totenopfern." *Archiv für Buchgewerbe und Gebrauchsgraphik*
79 (June): 221-222.

432 _____. 1939. "Stammt das Älteste Chinesische Papier vom Jahre
150 N. Chr.?" *Wochenblatt für Papierfabrikation* 70 (March
18): 244.

433 ———. 1940. "Sven Hedins Funde Ältesten Chinesischen Papiers." *Archiv für Buchgewerbe und Gebrauchsgraphik* 77 (January): 29.

434 Bojesen, C.C., and Rewi Alley. 1938. "China's Rural Paper Industry." *The China Journal* 28 (May): 233-243.

435 Chavannes, Édouard. 1905. "Les Livres Chinois Avant l'Invention du Papier." *Journal Asiatique (Tenth Series)* 5 (January-February): 5-75.

436 Chih-Hsiung, Deng. 1954. "Chinesische 'Reispapier' aus Holzmark." *Papiergeschichte* 4 (February): 9-12.

437 Friedemann, E. 1931. "Wann Hat der Erfinder des Papiers Gelebt?" *Der Papier-Fabrikant* 29 (August 2): 526-527.

438 Hunter, Dard. 1937. *Chinese Ceremonial Paper; A Monograph Relating to the Fabrication of Paper and Tin Foil and the Use of Paper in Chinese Rites and Religious Ceremonies.* Chillicothe, Ohio: The Mountain House Press.

439 ———. 1936. "Die Frühe Herstellung von Papier in China und Japan." *Zeitschrift für Büchfreunde (New Series)* 5 (March): 73-79.

440 ———. 1932. *Old Papermaking in China and Japan.* Chillicothe, Ohio: The Mountain House Press.

441 ———. 1936. *A Papermaking Pilgrimage to Japan, Korea and China.* New York: Pynson Printers.

442 Kan, Lao. 1948. "Lun Chung-kuo Tsao-chih-shu Ti Yüan-shih." *Bulletin of the Institute of History and Philology, Academia Sinica, Nanking and Taipei* 19: 496-498.

443 Laufer, Berthold. 1931. *Paper and Printing in Ancient China.* Chicago, Illinois: The Caxton Club.

444 ———. 1934. "Papier und Druck im Alten China." *Imprimatur* 5: 65-75.

445 Li, Shu-Hua. 1960. *The Spread of the Art of Paper-Making and the Discoveries of Old Paper.* Taipei, Taiwan: Collected Papers on History and Art of China, No. 3, National Historical Museum, Republic of China.

446 Narita, Kiyofusa. 1965. "A Life of Ts'ai Lun." *The Paper Maker* 34 (March): 18-27.

447 ———. 1972. "Vida de Ts'ai Lung, Inventor del Papel." *Investigación y Técnica del Papel* 9 (January): 97-105.

448 Pels, C. 1948. "Hoe Zij Hun Papier Maakten." *De Papierwereld* 2 (February): 608-609.

449 Renker, Armin. 1936. "Papier und Druck im Fernen Osten." *Der Papier-Fabrikant* 34 (June 28): 212-226.

450 _____. 1936. *Papier und Druck im Fernen Osten.* Mainz, Germany: Verlag der Gutenberg-Gesellschaft.

451 Schulte, Toni. 1956. "Kleider aus Papier im Fernen Osten." *Papiergeschichte* 6 (July): 43-44.

452 Seibert, Heinrich. 1971. "Vom China-Papier." *Papiergeschichte* 21 (December): 53-54.

453 Siu-ming, Chang. 1959. "A Note on the Date of the Invention of Paper in China." *Papiergeschichte* 9 (November): 51-52.

454 Tschichold, Jan. 1958. "Geschichte des Chinesischen Brief- und Gedrichtpapiers. Ein Beitrag Zur Geschichte und Zur Technik des Chinesischen Farbendruckes." *Philobiblon (Hamburg)* 2 (March): 31-56.

455 Tschudin, W. Fritz. 1954. "Quellen Zur Frühgeschichte des Papiers." *Textil-Rundschau* 9 (May): 244-251.

456 _____. 1952. "Über die Halzschnitt-Porträts Ts'ai Luns." *Papiergeschichte* 2 (April): 22-23.

457 Winczakiewicz, Andrzej. 1957. "Tsai-Lun Wynalazca Papieru." *Przegląd Papierniczy* 13 (April): 124-126.

458 Yui, Chien Hsuin. 1952. "Nachtrag Zur 'Studie der Geschichte des Chinesischen Papiers'." *Papiergeschichte* 2 (February): 5-6.

459 _____. 1951. "Studie Zur Geschichte des Chinesischen Papiers." *Papiergeschichte* 1 (October): 35-38.

EGYPT

460 Bockwitz, Hans H. 1952. "Zur Siebgrösse in der Altislamischen Papiermacherei Ägyptens." *Gutenberg-Jahrbuch* 27: 20.

461 Černý, Jaroslav. 1952. *Paper & Books in Ancient Egypt.* London, England: H.K. Lewis & Company.

462 Irigoin, Jean. 1963. "Les Types de Formes Utilisés dans l'Orient Mediterranéen (Syrie, Egypte) du XIe au XIVe Siècle." *Papiergeschichte* 13 (April): 18-21.

463 Lewis, N. 1934. *L'Industrie du Papyrus dans l'Egypte Gréco-Romaine.* Paris, France: Librarie L. Rodstein.

INDIA

464 Boatwala, Mini, and Wilfred Maciel. 1964. "Handmade Paper in India." *The Penrose Annual* 57: 281-283.

465 Bockwitz, Hans H. 1939. "Hat die Orientalische Papiermacherei Bereits Mechanische Stampfwerke Gekannt?" *Wochenblatt für Papierfabrikation* 70 (April 22): 349-352.

466 Chaudhary, Yadaviao S. 1936. *Handmade Paper in India.* Lucknow, India: J.C. Kumarappa on behalf of A.I.V.I.A.

467 Dutt, Asoka K. 1955. "Papermaking in India: a Resumé of the Industry From the Earliest Period Until the Year 1949." *The Paper Maker* 24 (September): 11-19.

468 Gode, P.K. 1944. "Migration of Paper From China to India." In *Paper Making As a Cottage Industry,* edited by K.B. Joshi. Wardha, India: V.L. Mehta. 4th Edition. 205-222.

469 Horne, C. 1877. "Paper Making in the Himalayas." *Indian Antiquary* 6 (April): 94-98.

470 Hunter, Dard. 1939. *Papermaking By Hand in India.* New York: Pynson Printers.

471 Lakshmi, R. 1957. "Handmade Paper in India." *The Paper Maker* 26 (September): 31-37.

472 Narayanswami, C.K. 1961. *The Story of Handmade Paper Industry.* Bombay, India: Khadi and Village Industries Commission. 2nd Edition.

473 Sandermann, Wilhelm. 1968. "Alte Techniken der Papierherstellung in Sudostasien und den Himalaya-Ländern." *Papiergeschichte* 18 (July): 29-39.

INDO-CHINA

474 Hunter, Dard. 1947. *Papermaking in Indo-China.* Chillicothe, Ohio: The Mountain House Press.

INDONESIA

475 Voorn, Henk. 1966. "Batik Paper." *The Paper Maker* 35 (June): 5-7.

476 _____. 1969. "Deloewang of Javaans Papier." *De Papierwereld* 24 (April): 95-99.

477 _____. 1968. "Javanese Deloewang Paper." *The Paper Maker* 37 (September): 32-38.

JAPAN

478 Bockwitz, Hans H. 1939. "Die Beiden Ältesten Chinesischen und Japanischen Papiermacher-Handbücher." *Wochenblatt für Papierfabrikation* 70 (February 25): 175-176.

479 ————. 1941. "Fernöstliche Papiermacherei der Gegenwart." *Archiv für Buchgewerbe und Gebrauchsgraphik* 78 (July): 246-252.

480 ————. 1939. "Hat die Orientalische Papiermacherei Bereits Mechanische Stampfwerke Gekannt?" *Wochenblatt für Papierfabrikation* 70 (April 22): 349-352.

481 Drissler, Hans. 1958. "Die Herstellung der Echten Japanpapiere." In *Probleme der Archivtechnik,* edited by Edgar Krausen. Munich, Germany: Bayer. Hauptstaatsarchiv. 47-50.

482 Hunter, Dard. 1936. "Die Frühe Herstellung von Papier in China und Japan." *Zeitschrift für Büchfreunde (New Series)* 5 (March): 73-79.

483 ————. 1932. *Old Papermaking in China and Japan.* Chillicothe, Ohio: The Mountain House Press.

484 ————. 1936. *A Papermaking Pilgrimage to Japan, Korea and China.* New York: Pynson Printers.

485 Jugaku, Bunshō. 1959. *Papermaking By Hand in Japan.* Tokyo, Japan: Meiji-Shobo.

486 ————. 1957. "Where They Still Make Paper By Hand." *Japan Quarterly* 4 (April-June): 249-251.

487 Narita, Kiyofusa. 1965. "A Brief History of Papermaking By Hand in Japan." *The Paper Maker* 34 (September): 5-14.

488 ————. 1953. "Die Einführung der Industrie und der Modernen Papierfabrikation in Japan." *Papiergeschichte* 3 (July): 42-43.

489 ————. 1957. "The First Machine-Made Paper in Japan." *The Paper Maker* 26 (February): 11-15.

490 ————. 1959. "Japanese Paper and Paper Products." *Papiergeschichte* 9 (December): 84-87.

491 ————. 1953. "Japanische Schöpfformen in Alter und Neuer Zeit." *Papiergeschichte* 3 (July): 36-37.

492 ————. 1962. "Making Paper By Hand in Japan." *The Paper Maker* 31 (March): 38-58.

493 ————. 1972. "Sucinta Historia del Papel Japonés Fabrícado a Mano." *Investigación y Técnica del Papel* 9 (April): 465-477.

494 ————. 1955. "Suminagashi." *The Paper Maker* 24 (February): 27-31.

495 ————. 1973. "El 'Suminagashi', Papel Pintado Por Flotación en Tinta o Papel Jaspeado." *Investigación y Técnica del Papel* 10 (April): 487-492.

496 Nissen, C. 1933. "Über die Verfertigung des Papiers in Japan." *Wochenblatt für Papierfabrikation* 64 (Special Edition): 13-15.

497 Reifegerste. 1929. "Handpapiermacherei in Japan." *Wochenblatt für Papierfabrikation* 60 (June 8A): 20-33.

498 Stevens, Richard Tracy. 1909. *The Art of Papermaking in Japan.* New York: Privately Printed.

499 Tindale, Thomas Keith, and Harriet Ramsey Tindale. 1952. *The Handmade Papers of Japan.* Rutland, Vermont: Charles E. Tuttle Company.

500 Voorn, Henk. 1950. "Een Herdruk van Het Oudste Japanse Papiermakershandboek in Voorbereiding." *De Papierwereld* 4 (July): 440.

KOREA

501 Bockwitz, Hans H. 1941. "Fernöstliche Papiermacherei der Gegenwart." *Archiv für Buchgewerbe und Gebrauchsgraphik* 78 (July): 246-252.

502 ————. 1939. "Hat die Orientalische Papiermacherei Bereits Mechanische Stampfwerke Gekannt?" *Wochenblatt für Papierfabrikation* 70 (April 22): 349-352.

503 Hunter, Dard. 1936. *A Papermaking Pilgrimage to Japan, Korea and China.* New York: Pynson Printers.

THE MIDDLE EAST

504 Bockwitz, Hans H. 1951. "Die Früheste Verwendung von Papier in den Altislamischen Kanzleien." *Papiergeschichte* 1 (October): 39-40.

505 ————. 1941. "Das Papier im Alten Weltreich des Islam." *Archiv für Buchgewerbe und Gebrauchsgraphik* 78 (August): 309-312.

506 ————. 1955. "Ein Papierfund aus dem Anfang des 8. Jahrhunderts am Berge Mugh Bei Samarkand." *Papiergeschichte* 5 (July): 42-44.

507 Garnett, R. 1903. "Early Arabian Paper Making." *The Library (Second Series)* 4 (January): 1-10.

508 Karabacek, Joseph. 1887. "Das Arabische Papier." *Mittheilungen aus der Sammlung der Papyrus Erzherzog Rainer* 2-3: 114.

509   Voorn, Henk. 1950. "De Arabische Periode in de Geschiedenis van Het Papier. I. Samarkand." *De Papierwereld* 4 (March): 290-291.

510   _____. 1950. "De Arabische Periode in de Geschiedenis van Het Papier. II. Van Samarkand Tot Fez." *De Papierwereld* 4 (March): 316.

511   _____. 1950. "De Arabische Periode in de Geschiedenis van Het Papier. III. Katoenpapier." *De Papierwereld* 4 (April): 330-332.

512   _____. 1950. "De Arabische Periode in de Geschiedenis van Het Papier. IV. De Papiermaker." *De Papierwereld* 4 (April): 350-353.

513   _____. 1950. "De Arabische Periode in de Geschiedenis van Het Papier. V. Arabisch Papier." *De Papierwereld* 4 (May): 370-371.

514   _____. 1959. "Papermaking in the Moslem World." *The Paper Maker* 28 (February): 31-38.

515   Weir, Thomas S. 1957. "Some Notes on the History of Papermaking in the Middle East." *Papiergeschichte* 7 (July): 43-48.

NEPAL

516   Beatty, W.B. 1962. "The Handmade Paper of Nepal." *The Paper Maker* 31 (September): 13-25.

517   Sen, Siva Narayana. 1940. "Hand-Made Paper of Nepal." *The Modern Review* 67 (April): 459-463.

518   Trier, Jesper. 1972. *Ancient Paper of Nepal.* Copenhagen, Denmark: Jutland Archaeological Society Publications.

THE SOUTH SEA ISLANDS

519   Hunter, Dard. 1926. "Making Paper in the South Sea Islands." *The American Printer* 83 (September): 33-36.

520   _____. 1927. "Paper Making in the South Seas." *Pulp and Paper Magazine of Canada* 25 (May 5): 580.

SYRIA

521   Irigoin, Jean. 1963. "Les Types de Formes Utilisés dans l'Orient Mediterranéen (Syrie, Égypte) du XIe au XIVe Siècle." *Papiergeschichte* 13 (April): 18-21.

THAILAND

522 Hunter, Dard. 1936. *Papermaking in Southern Siam.* Chillicothe, Ohio: The Mountain House Press.

TIBET

523 Nebesky-Wojkowitz, R. 1949. "Schriftwesen, Papierherstellung und Buchdruck Bei den Tibetein." Unpublished Ph.D. dissertation, University of Wien.

TURKEY

524 Babinger, Franz. 1931. "Appunti Sulle Cartiere e Sull'Importazione di Carta Nell'Impero Ottomano Specialmente da Venezia." *Oriente Moderne* 11 (August): 406-415.

525 ————. 1931. *Zur Geschichte der Papiererzeugung im Osmanischen Reiche.* Berlin, Germany: Reichsdruckerei.

526 Bockwitz, Hans H. 1948. "Hochachtung Vor dem Papier Bei den Türken im 16. Jahrhundert." *Wochenblatt für Papierfabrikation* 76 (January): 12-13.

527 Ersoy, Osman. 1963. *XVIII. Ve XIX. Yüzyillarda Türkiye'de Kâğit.* Ankara, Turkey: Ankara Üniversitesi Dil ve Tarih-Coğrafya Fakültesi, Yayimari-145.

528 Froundjian, Diraér. 1967. "Ein Armenisches Papier aus dem Jahre 1641." *Papiergeschichte* 17 (May): 23-24.

529 ————. 1964. "Die Erste Papiermühle in Armenien." *Papiergeschichte* 14 (July): 28-29.

530 Kağitçi, Mehmed Ali. 1963. "Beitrag Zur Türkischen Papiergeschichte." *Papiergeschichte* 13 (November): 37-44.

531 ————. 1965. "A Brief History of Papermaking in Turkey." *The Paper Maker* 34 (September): 44-51.

532 Mošin, V., and M. Grozdanović-Pajić. 1963. "Das Wasserzeichen 'Krone Mit Stein und Halbmond'." *Papiergeschichte* 13 (November): 44-52.

533 Nikolaev, Vsevolod. 1954. *Watermarks of the Ottoman Empire.* Sofia, Bulgaria: Bulgarian Academy of Sciences.

534 Schulte, Toni. 1958. "Kleine Hindweise Zur Papierfabrikation in der Türkei." *Papiergeschichte* 8 (July): 43-48.

535 ————. 1958. "Kleine Hindweise Zur Papierfabrikation in der Türkei." *Papiergeschichte* 8 (September): 50-52.

VIETNAM

536   Bockwitz, Hans H. 1939. "Bilder aus dem Papiermacherdorf Hanoi in Tonkin." *Wochenblatt für Papierfabrikation* 70 (August 26): 747-748.

3   *The History of*
Paper and Papermaking
in Europe
and the Soviet Union

AUSTRIA

537 Bogdán, István. 1962. "Egy 'Elégtelen' Papírtörténeti Kiaclványrol." *Papíripar es Magyar Grafika* 6 (January-February): 45-46.
538 Eineder, Georg. 1960. *The Ancient Paper-Mills of the Former Austro-Hungarian Empire and Their Watermarks.* Hilversum, The Netherlands: The Paper Publications Society.

BELGIUM

539 Bock, R. de. 1946. "Een Handel in Pennen en Papier te Antwerpen Einde XVIIIe, Begin 19e Eeuw." *De Gulden Passer* 24: 1-17.
540 Godenne, Willy. 1961. "Trois Textes Originaux Concernant la Papeterie en Belgique au Début du XVIII°s." *Papiergeschichte* 11 (October): 46-51.
541 Tacke, Eberhard. 1958. "Die Belgische Papierindustrie Nach Einem Reisebericht vom Jahre 1841." *Papiergeschichte* 8 (February): 15.
542 Voorn, Henk. 1953. "Drie-en-een-halve Eeuw Papiermaker in Saventhem." *De Papierwereld* 8 (August): 18-19.

CZECHOSLOVAKIA

543 Čarek, J. 1936. "Památky Choceňské Papirny." *Časopis Společnosti Přátel Starožitnosti* 44: 182-184.
544 Decker, Viliam. 1960. "Die Anfänge und die Entwicklung des Papiergebrauchs in der Slowakei." *Papiergeschichte* 10 (July): 35-36.
545 _____. 1958. "Entwurf der Geschichte der Hand-Papiererzeugung in der Slowakei." *Papiergeschichte* 8 (December): 81-82.
546 _____. 1963. "Papieren v Ochtinej." *Papír a Celulósa* 18 (November): 225-227.
547 _____. 1962. "Priesvitky Papierni v Meste Kremnici." *Papír a Celulósa* 17 (November): 255-257.
548 Fiala, Josef. 1949. "Papírna v Zámku Zdáre na Morave." *Papír a Celulósa* 4: 15-17.
549 Florian, Č. 1940. "Papirna ve Svidnici u Chrudimě." *Časopis Národniho Musea* 114: 73-81.
550 Gerold, Vladimir. 1958. "Filigrány Staré Opavské Papirny." *Papír a Celulósa* 13 (August): 181-183.

41

551 ————. 1961. "Papírna ve Velkých Losinách." *Papír a Celulósa* 16 (July): 158-160.

552 ————. 1957. "Staré Opavské Papírny." *Papír a Celulósa* 12: 283-284.

553 Hekele, F. 1958. "Z Historie Moravských Papíren - Papírna v Přibyslavicích." *Papír a Celulósa* 13: 207.

554 ————. 1958. "Z Minulosti Moravských Papiren." *Papír a Celulósa* 13: 285.

555 ————. 1958. "Staré Moravské Papírny." *Rodné Zemi. Sborník k 70. Výroří Tráni Musejního Spolku v Brně* n.v.: 160-171.

556 ————. 1957. "Zaniklá Papírna na Mirově." *Severni Morava* 2: 55-56.

557 ————. 1957. "Zaniklé Severomoravské Papírny." *Vlastivědný Věstnik Moravský* 12: 230-234.

558 ————. 1957. "Ze Života Starých Výrobcu Papirû." *Papír a Celulósa* 12: 179-180.

559 Korda, Josef. 1972. "Počátky Papíru, Tisku a Autorských Práv Čechách." *Papír a Celulósa* 27 (February): 39-40.

560 Meissner, Frank. 1960. "Some Notes About a Papermakers' Association in Slovakia During the Eighteenth Century." *The Paper Maker* 29 (September): 43-48.

561 Schusser, František. 1974. "Z Historie Papíren Vlavský Mlýn." *Papír a Celulósa* 29 (October): 234-237.

562 Šilhan, J. 1960. "Papírna v Dlouhé Loučce do Poloviny 18. Stol." *Časopis Společnosti Přátel Starožitnosti* 68: 83-89.

563 Slíva, Oldřich. 1974. "Minulost Současnost a Budoucnost Vratimovských Papíren." *Papír a Celulósa* 29 (April): 81-82.

564 Tywoniak, J. 1955. "Konopištská Papírna (Konec 18. Stol.-1851)." *Časopis Společnosti Přátel Starožitnosti* 63: 175-187.

565 Zuman, František. 1923. "České Filigrány XVI Století." *Památky Archeologické* 33: 277-286.

566 ————. 1927. "České Filigrány XVII Století." *Památky Archeologické* 35: 442-463.

567 ————. 1929-1930. "České Filigrány XVII Století. II." *Památky Archeologické* 36: 268-270.

568 ————. 1931. *České Filigrány XVIII Století.* Prague, Czechoslovakia: Rozprazy České Akademie Věd a Umění.

569 ————. 1934. *České Filigrány z První Polovice XIX Století.* Prague, Czechoslovakia: Rozpravy České Akademie Věd a Umění.

570 _____. 1952. "Co Víme o První České Papírně." *Časopis Společnosti Přátel Starožitností* 60: 176-178.

571 _____. 1922. "Eiligrán Mimoňský a Hamerský." *Památky Archeologické* 33: 158-160.

572 _____. 1939. "Filigrány, Jejich Vývoj a Význam." *Casopis Národního Musea* 113: 88-117.

573 _____. 1922. "Inventáře Papírny z Poč 17. Století." *Památky Archeologické* 33: 167-168.

574 _____. 1923. "Inventáře Bělské Papírny z r. 1723." *Památky Archeologické* 33: 344-345.

575 _____. 1947. *Knízka o Papíru.* Prague, Czechoslovakia: Společnost Přátel Starožitností.

576 _____. 1947. "Lužnicka Papírna v Šerachově." *Papír a Celulósa* 2 (September): 19.

577 _____. 1940. "Nové České a Moravské Papírnické Obchodní Značky." *Časopis Národního Musea* 114: 58-66.

578 _____. 1937. "Nově Objevené Obchodní Značky Papírnické." *Časopis Národního Musea* 111: 105-111.

579 _____. 1936. "Die Papiermühle in Schirgiswalde. Ein Beitrag Zur Geschichte Derselben aus den Prager Archiven." *Der Altenburger Papierer* 10: 750-755.

580 _____. 1948. "Papír a Heraldika (Studie Zapomenutých Pramenů)." *Rodokmen* 3: 73-80.

581 _____. 1949. "Papírna Augustinianského Řádu v Pivoni." *Papír a Celulósa* 4 (November): 17-23.

582 _____. 1932. "Papírna Telnická." *Časopis Společnosti Přátel Starožitností* 40: 21-31.

583 _____. 1933. "Papírna v Bělé Pod Bezdězem." *Od Ještěda k Troskám* 11: 173-182.

584 _____. 1932. "Papírna v Borečku u Kumru." *Bezděz* 3: 164-177.

585 _____. 1950. "Papírna v Březinách u Děcína." *Papír a Celulósa* 5: 16.

586 _____. 1951. "Papírna v Bubenči." *Časopis Společnosti Přátel Starožitností* 59: 179-183.

587 _____. 1950. "Papírna y Dobrovicích u Tábora." *Papír a Celulósa* 5 (December): 163.

588 _____. 1934. "Papírna v Dolní Polici." *Bezděz* 5: 63-88.

589 _____. 1950. "Papírna v Dolní Poustevně." *Papír a Celulósa* 5 (September): 130-132.

590 _____. 1949. "Papírna v Dolním Nýrsku." *Papír a Celulósa* 4 (September): 19-20.

591 _____. 1931. "Papírna v Dubi na Teplickém Panstvi." *Časopis Společnosti Přátel Starožitnosti* 39: 84-94.

592 _____. 1948. "Papírna v Janovic." *Papír a Celulósa* 3 (November): 19.

593 _____. 1950. "Papírna v Michalových Horách." *Papír a Celulósa* 5 (December): 177-178.

594 _____. 1921. "Papírna v Pardubicich." *Časopis Musea Královstvi Českého* 95: 139-148.

595 _____. 1948. "Papírna v Postřekově na Chodsku." *Papír a Celulósa* 3 (June): 21-22.

596 _____. 1921. "Papírna v Prášilech." *Zlatá Praha* 38: 183-185.

597 _____. 1951. "Papírna v Předklášteří u Tišnova." *Vlastivedný Věstnik Moravský* 6: 47-56.

598 _____. 1947. "Papírna v Rádlu." *Papír a Celulósa* 2 (November): 23-24.

599 _____. 1935. "Papírna Žačléřská." *Časopis Společnosti Přátel Starožitnosti* 43: 110-124.

600 _____. 1950. "Papírna v Zámrsku u Vys Mýta." *Papír a Celulósa* 5 (June): 75.

601 _____. 1927. "Papírny na Panstvi Lilomyšlském." *Časopis Společnosti Přátel Starožitnosti* 35: 150-160.

602 _____. 1949. "Papírny na Svitavě." *Časopis Společnosti Přátel Starožitnosti* 57: 227-236.

603 _____. 1947. "Papírny 16., 17., 18. a 19. Stol. v Čechách." *Papír a Celulósa* 2: 25-26.

604 _____. 1933. "Počátky Papírny v Benešově Nad Ploučniki." *Bezděz* 4: 182-185.

605 _____. 1934. *Pootavské Papírny*. Prague, Czechoslovakia: Vydal Archiv Pro Dějiny Průmyslu, Obchodu Technické Práce v Praze.

606 _____. 1936. *Posázavské Papírny*. Prague, Czechoslovakia: Vydal Archiv Pro Dějiny Průmyslu, Obchodu Technické Práce v Praze.

607 _____. 1947-1948. "Posledni Doby Vrchlabské Papírny." *Horské Prameny Vlastivedný Sborník Krkonoš* 3: 52-55.

608 _____. 1947-1948. "Posledni Doby Vrchlabské Papírny." *Horské Prameny Vlastivedný Sborník Krkonoš* 3: 105-109.

609 _____. 1934. "Přehled Papíren v Čechách v 1. Polovici 19. Stoleti." *Časopis Národního Musea* 108: 26-52, 197-224.

610 _____. 1921. "Přehled Papíren v Čechách v 17. Stoleti." *Český Časopis Historický* 27: 162-170.

611 _____. 1931. "Přehled Papíren v Čechách v 18. Století." *Český Časopis Historický* 37: 79-90, 293-309.

612 _____. 1935. "Rumburské Výrobky z Papíroviny." *Bezděz* 6: 40-44.

613 _____. 1923. "Stará Velhartická Papírna." *Zlatá Praha* 40: 197-198.

614 _____. 1947. "Stodeset Let Papírny v Vraném." *Papír a Celulósa* 2 (October): 17-18.

615 _____. 1921. "Vodní Značka Průvodním Prostředkem." *Památky Archeologické* 32: 260-261.

616 _____. 1937. *Výrobní Technika Papíru a Její Vývoj.* Prague, Czechoslovakia: Vydal Archiv Pro Dějiny Průmyslu, Technické Práce v Praze.

617 _____. 1936. "Vzácná Obchodníznačka Papírny." *Časopis Národního Musea* 110: 63-73.

618 _____. 1948. "Zaniklá Papírna v Sorgentálu." *Papír a Celulósa* 3 (January): 18-19.

619 _____. 1939. "Das Zeremoniell der Friesprechung im 19. Jahrhundert." *Der Altenburger Papierer* 13: 287-289.

620 _____. 1937. "Zlořády v Papírnách Našeho Území." *Bezděz* 8: 18-19.

621 _____. 1930. "Zlořády v Řemesle Papírnickém." *Časopis Národního Musea* 104: 234-250.

DENMARK

622 Nordstrand, Ove K. 1962. "On the Preparation of a Danish Water-Mark Collection." *Papiergeschichte* 12 (February): 18-20.

623 Voorn, Henk. 1955. "Papermaking in Denmark." *The Paper Maker* 24 (February): 1-17.

624 _____. 1959. *The Paper Mills of Denmark & Norway and Their Watermarks.* Hilversum, The Netherlands: The Paper Publications Society.

ENGLAND

625 Balston, Thomas. 1957. *James Whatman, Father and Son.* London, England: Methuen.

626 _____. 1959. "Whatman Paper in a Book Dated 1757." *The Book Collector* 8 (Autumn): 306-308.

627 _____. 1954. *William Balston, Paper Maker, 1759-1849.* London, England: Methuen.

628 Bockwitz, Hans H. 1938. "Zur Geschichte der Ältesten Englischen Papiermacherei." *Wochenblatt für Papierfabrikation* 69 (August 13): 686-687.

629 _____. 1938. "Ein Illustrierter Englischer. Konsultatsbericht Über Japanische Papiermacherei vom Jahre 1870." *Wochenblatt für Papierfabrikation* 69 (February 12): 156-157.

630 _____. 1942. "Johann Spielmann, ein Deutscher Papiermacher des 16. Jahrhunderts in England." *Archiv für Buchgewerbe und Gebrauchsgraphik* 79: 41.

631 _____. 1942. "Johann Spielmann, ein Deutscher Papiermacher in England im 16. Jahrhundert." *Wochenblatt für Papierfabrikation* 73 (April 18): 119.

632 _____. 1939. "Willcox-Papier." *Wochenblatt für Papierfabrikation* 70 (March 18): 244.

633 Bridge, W.E. 1948. *Some Historical Notes on the Basted Paper Mills.* Kent, England: Basted Paper Mills Company.

634 Cameron, W.J. 1964. *The Company of White-Paper-Makers of England 1686-1696.* Auckland, New Zealand: Bulletin 68, Economic History Series 1, University of Auckland.

635 Carter, Harry. 1957. *Wolvercote Mill: a Study in Paper-Making at Oxford.* Oxford, England: The Clarendon Press.

636 Chapman, R.W. 1927. "An Inventory of Paper, 1674." *The Library (Fourth Series)* 7 (March): 402-408.

637 Churchill, W.A. 1935. *Watermarks in Paper in Holland, England, France, etc., in the XVII and XVIII Centuries and Their Interconnection.* Amsterdam, Holland: M. Hertzberger & Company.

638 Clapperton, R.H. 1953. "The History of Papermaking in England." *The Paper Maker* 22 (September): 9-22.

639 Coleman, D.C. 1958. *The British Paper Industry 1495-1860: a Study in Industrial Growth.* Oxford, England: The Clarendon Press.

640 _____. 1954. "Combinations of Capital and Labour in The English Paper Industry 1789-1825." *Economica (New Series)* 21 (February): 32-53.

641 _____. 1956. "Industrial Growth and Industrial Revolutions." *Economica (New Series)* 23 (February): 1-22.

642 Cornett, J.P. 1908. "Local Paper Mills." *Antiquities of Sunderland and Its Vicinity* 9: 162-167.

643   Elliott, Harrison. 1951. "John Baskerville (1706-1775): Type-Founder, Typographer, Printer and Originator of Wove Paper." *The Paper Maker* 20 (September): 11-15.

644   _____. 1953. "The Portals: Papermakers to the Bark of England." *The Paper Maker* 22 (February): 38-42.

645   Evans, Lewis. 1896. *The Firm of John Dickinson and Company Limited.* London, England: Chiswick Press.

646   Gaskell, Philip. 1957. "Notes on Eighteenth-Century British Paper." *The Library (Fifth Series)* 12 (March): 34-42.

647   George, Robert H. 1931. "A Mercantilist Episode." *Journal of Economic and Business History* 3: 264-271.

648   Gibson, J.R. 1958. "Paper Industry of North-West England." Unpublished Masters thesis, University of Liverpool.

649   Hazen, Allen T. 1952-1953. "Baskerville and James Whatman." *Studies in Bibliography* 5: 187-189.

650   _____. 1954. "Eustace Burnaby's Manufacture of White Paper in England." *The Papers of the Bibliographical Society of America* 48 (Fourth Quarter): 315-333.

651   Heawood, Edward. 1947. "Further Notes on Paper Used in England After 1600." *The Library (Fifth Series)* 2 (September): 119-149.

652   _____. 1930. "Papers Used in England After 1600: I. the Seventeenth Century to c. 1680." *The Library (Fourth Series)* 11 (December): 263-299.

653   _____. 1931. "Papers Used in England After 1600: II. c. 1680-1750." *The Library (Fourth Series)* 11 (March): 466-498.

654   _____. 1948. "Papers Used in England After 1600 (Correspondence)." *The Library (Fifth Series)* 3 (September): 141-142.

655   _____. 1929. "Sources of Early English Paper-Supply." *The Library (Fourth Series)* 10 (December): 282-307.

656   _____. 1930. "Sources of Early English Paper-Supply: II. the Sixteenth Century." *The Library (Fourth Series)* 10 (March): 427-454.

657   Hernlund, Patricia. 1969. "William Strahan's Ledgers, II: Charges for Papers, 1738-1785." *Studies in Bibliography* 22: 179-195.

658   Hössle, F. von. 1930. "Der Goldschmied und Papiermacher Spielmann." *Wochenblatt für Papierfabrikation* 61 (June 21A): 11-17.

659   Jenkins, Rhys. 1900. "Early Attempts at Paper-Making in England, 1495-1586." *The Library Association Record* 2 (September): 479-488.

660 _____. 1914. "Paper-Making in Devon." *Devon & Cornwall Notes & Queries* 8 (October): 119-121.

661 _____. 1936. "Paper-Making in England, 1495-1788." In *The Collected Papers of Rhys Jenkins,* by Rhys Jenkins. Cambridge, England: Printed for the Newcomen Society at the University Press. 155-192.

662 _____. 1900. "Paper-Making in England, 1588-1680." *The Library Association Record* 2 (November): 577-588.

663 _____. 1901. "Paper-Making in England, 1682-1714." *The Library Association Record* 3 (May): 239-251.

664 _____. 1902. "Paper-Making in England, 1714-1788." *The Library Association Record* 4 (March-April): 128-139.

665 Kunze, Horst. 1941. "Matthias Koops und die Erfindung des Strohpapiers." *Gutenberg-Jahrbuch* 16: 30-45.

666 Labarre, E.J. 1952. "The Study of Watermarks in Great Britain." In *The Briquet Album,* edited by E.J. Lararre. Hilversum, The Netherlands: The Paper Publications Society. 97-106.

667 LaRue, Jan. 1957. "British Music Paper 1770-1820: Some Distinctive Characteristics." *Monthly Musical Record* 87 (September-October): 177-180.

668 Lega-Weekes, Ethel. 1920. "Early Use of Paper in Devonshire: Watermarks." *Devon & Cornwall Notes & Queries* 11 (July): 108-110.

669 Lewis, Peter W. 1967. "Changing Factors of Location in the Paper-making Industry As Illustrated By the Maidstone Area." *Geography* 52 (July): 280-293.

670 _____. 1968. "A Geography of the Papermaking Industry in England and Wales, 1860-1965." Unpublished Ph.D. dissertation, Manchester University.

671 _____. 1969. *A Numerical Approach to the Location of Industry.* Hull, England: University of Hull Publications.

672 Lloyd, L.C. 1937-1938. "Paper-Making in Shropshire, 1656-1912: Some Records of a Byegone Industry." *Transactions of the Shropshire Archaeological and Natural History Society* 49: 121-187.

673 _____. 1950. Paper-Making in Shropshire: Supplementary Notes." *Transactions of the Shropshire Archaeological and Natural History Society* 53: 153-163.

674 Lopez, R. 1940. "The English and the Manufacture of Writing Materials in Genoa." *Economic History Review* 10 (November): 132-137.

675 Macfarlane, John. 1899. "The Paper Duties of 1696-1713; Their Effect on the Printing and Allied Trades." *The Library (Second Series)* 1 (December): 31-44.

676 Mason, John. 1963. "Adventurous Papermaking: the Founding of the Twelve By Eight Mill." *The Black Art* 2 (Autumn): 74-78.

677 Oldman, C.B. 1944. "Watermark Dates in English Paper." *The Library (Fourth Series)* 25 (June-September): 70-71.

678 Overend, G.H. 1909. "Notes Upon the Earlier History of the Manufacture of Paper in England." *Proceedings of the Huguenot Society of London* 8: 177-220.

679 Parris, H. 1960. "Adaption to Technical Change in the Paper-Making Industry: the Paper-Mill at Richmond, Yorkshire, 1823-1846." *Yorkshire Bulletin of Economic and Social Research* 12 (November): 84-89.

680 Pelham, R.A. 1945-1946. "Hutton's Paper Mill and Its Geographical Significance." *Transactions and Proceedings of the Birmingham Archeological Society* 66: 155.

681 Richards, P.S. 1965. "The Siting of Two Flintshire Paper Mills." *Journal of Industrial Archaeology* 2 (December): 161-166.

682 Richardson, R. Morris. 1908. "The Hutton Family (Papermakers)." *Antiquities of Sunderland and Its Vicinity* 9: 168-179.

683 Riddle, E.C. 1907. *Whatman Paper Mill: the Centenary of Springfield Mill, Maidstone.* Maidstone Kent, England: Springfield Mill.

684 Shears, W.S. 1950. *William Nash of St. Paul's Cray; Papermakers.* London, England: Batchworth Press.

685 Shorter, Alfred H. 1938. "The Distribution of Paper-Making in Cornwall in the Nineteenth Century." *Devon & Cornwall Notes & Queries* 20 (January): 2-10.

686 _____. 1951. "Early Paper-Mills in Kent." *Notes & Queries* 196 (July 21): 309-313.

687 _____. 1950. "The Excise Numbers of Paper-Mills in Shropshire." *Transactions of the Shropshire Archaeological and Natural History Society* 53: 145-152.

688 _____. 1950. "The Historical Geography of the Paper-Making Inudstry in Devon, 1684-1950." *Transactions of the Devonshire Association for the Advancement of Science, Literature and Art* 82: 205-216.

689 _____. 1954. "The Historical Geography of the Paper Making Industry in England." Unpublished Ph.D. dissertation, University of London.

690 _____ . 1950. "Paper and Board Mills in Somerset." *Somerset and Dorset Notes & Queries* 25 (March): 245-256.

691 _____ . 1938. "Paper-Making in Devon and Cornwall." *Geography* 23 (September): 164-176.

692 _____ . 1938. "The Paper-Making Industry Near Barnstaple." *Devon & Cornwall Notes & Queries* 20 (April): 56-57.

693 _____ . 1972. *Paper Making in the British Isles: an Historical and Geographical Study.* New York: Barnes & Noble, Publishers.

694 _____ . 1957. *Paper Mills and Paper Makers in England, 1495-1800.* Hilversum, The Netherlands: The Paper Publications Society.

695 _____ . 1938. "Paper Mills in Devon." *Devon & Cornwall Notes & Queries* 20 (January): 34-35.

696 _____ . 1947. "Paper-Mills in Devon and Cornwall." *Devon & Cornwall Notes & Queries* 23 (October): 97-103.

697 _____ . 1951. "Paper-Mills in Devon and Cornwall." *Devon & Cornwall Notes & Queries* 24 (October): 221-222.

698 _____ . 1948. "Paper-Mills in Devon and Cornwall: Further Evidence." *Devon & Cornwall Notes & Queries* 23 (July): 193-198.

699 _____ . 1948. "Paper-Mills in Dorset." *Somerset and Dorset Notes & Queries* 25 (December): 144-148.

700 _____ . 1960. "Paper Mills in England in the 1820's." *Papiergeschichte* 10 (July): 32-35.

701 _____ . 1952. "Paper Mills in Gloucestershire." *Transactions of the Bristol and Gloucestershire Archaeological Society* 71.

702 _____ . 1953. "Paper Mills in Hampshire." *Proceedings of the Hampshire Field Club and Archaeological Society* 18.

703 _____ . 1951. "Paper Mills in Herefordshire." *Transactions of the Woolhope Naturalists' Field Club* 34.

704 _____ . 1953. "Paper-Mills in Monmouthshire." *Archaeologia Cambrensis* 102: 83-88.

705 _____ . 1952. "Paper-Mills in Somerset." *Somerset and Dorset Notes & Queries* 26 (April): 69.

706 _____ . 1950. "Paper-Mills in Somerset and Dorset: Further Evidence." *Somerset and Dorset Notes & Queries* 25 (September): 300-301.

707 _____ . 1951. "Paper Mills in Sussex." *Sussex Notes and Queries* 13.

708 _____ . 1951. "Paper Mills in Worcestershire." *Transactions of the Worcestershire Archaeological Society* 28.

709 _____. 1963. "Paper Mills—Their Geographical Distribution in Britain." *Paper & Print* 36 (Summer): 157-159, 162.

710 _____. 1966. *Water Paper Mills in England.* London, England: Society For the Protection of Ancient Buildings.

711 Sporhan-Krempel, Lore. 1963. "Zur Geschichte der Lindauer Papiermühlen." *Papiergeschichte* 13 (April): 4-7.

712 _____. 1963. "Hans Spilman From Lindau." *The Paper Maker* 32 (September): 16-20.

713 Stevenson, Allan H. 1951. "A Critical Study of Heawood's 'Watermarks Mainly of the 17th and 18th Centuries'." *The Papers of the Bibliographical Society of America* 45 (First Quarter): 23-36.

714 Sugden, Alan V., and John L. Edmondson. 1926. *A History of English Wallpaper, 1509-1914.* London, England: B.T. Batsford.

715 Sullivan, Frank. 1951. "Little Pitchers in the Big Years: Being a Study of the Water Pitcher Watermark in Elizabethan England." *The Paper Maker* 20 (February): 11-19.

716 Summers, W.H. 1894. "Early Paper Mills in Buckinghamshire." *Records of Buckinghamshire* 7.

717 Tann, Jennifer. 1967. "Multiple Mills." *Medieval Archaeology* 2: 253-255.

718 Voorn, Henk. 1967. "Anglo-Dutch Relations in Paper-Making History." *De Papierwereld* 22 (October): 293-296.

719 _____. 1969. "John Mathew, Papermaker." *The Paper Maker* 38 (June): 3-26.

720 _____. 1949. "Mathias Koops' 'Invention of Paper'." *De Papierwereld* 4 (October): 106-108.

721 Wardrop, James. 1938. "Mr. Whatman, Papermaker." *Signature* 9 (July): 1-18.

722 Watkin, Hugh R. 1920. "Early Use of Paper in Devonshire." *Devon & Cornwall Notes & Queries* 11 (January): 33-36.

723 _____. 1920. "Paper-Making in Devonshire." *Devon & Cornwall Notes & Queries* 11 (January): 33.

724 Watson, B.G. 1967. "John Dickinson and His Paper Machine." *The Paper Maker* 36 (September): 32-55.

725 _____. 1957. "The Search for Papermaking Fibers: Thomas Routledge and the Use of Esparto Grass As a Papermaking Fiber in Great Britain." *The Paper Maker* 26 (February): 1-6.

726 Wheelwright, William Bond. 1943. "Matthias Koops, Papermaker." *The Paper Maker* 12 (October): 3-5.

727   Wicks, A.T. 1950. "Paper Mill at Monkton Coombe." *Somerset and Dorset Notes & Queries* 25 (September): 296.

728   Williamson, Kenneth. 1939. "Paper Making in Man: the Story of a Lost Industry." *The Journal of the Manx Museum* 4 (June): 126-129.

729   _____. 1939. "Paper Making in Man: the Story of a Lost Industry." *The Journal of the Manx Museum* 4 (September): 147-151.

730   Wise, R.G. 1958. "The Historical and Social Background of Papermaking in England During Three Centuries." *The Paper Maker* 27 (September): 31-38.

FINLAND

731   Burgman, Oskar. 1955. "Ein Kurzer Bericht Über Einige Alte Papierfabriken in Finnland." *Papiergeschichte* 5 (December): 82-84.

732   Karlsson, Kurt K. 1963. "Die Erste Papiermühle in Finnland und Ihre Wasserzeichen." *Papiergeschichte* 13 (April): 11-18.

733   _____. 1962. "Sonderbriefmarke Anlässlich der 10. Nordischen Papieringenieurversammlung." *Papiergeschichte* 17 (November): 71-72.

734   _____. 1962. "Om Vattenmärken i Handgjort Papper i Finland." *Paperi Ja Puu* 44 (September): 459-472.

FRANCE

735   Alibaux, Henri. 1952. "Bibliographie d'Ouvrages Concernant l'Histoire du Papier et les Filigranes Antérieurement à 1600, Publiés en France et en Belgique Depuis 1907." In *The Briquet Album,* edited by E.J. Labarre. Hilversum, The Netherlands: The Paper Publications Society. 44-48.

736   _____. 1926. "The French Paper Industry in the XVIth, XVIIth and XVIIIth Centuries." *Paper Trade Journal* 54 (August 5): 59-60.

737   _____. 1926. "L'Industrie Française du Papier aux XVIe, XVIIe et XVIIIe Siècles." *Le Papier* 29 (April): 347-359.

738   _____. 1935. "Les 'Naypiers' et l'Origine des Cartes à Jouer." *Contribution à l'Histoire de la Papeterie en France* 2: 109-115.

739   _____. 1926. *Les Premières Papeteries Françaises.* Paris, France: Les Arts et le Livre.

740   Apcher, Louis. 1941. "A Propos du Saint, Patron des Fabricants de Papier." *Contribution à l'Histoire de la Papeterie en France* 7: 15-18.

741   Audin, Marius. 1938. "Les Anciennes Papeteries du Beaujolais et le Chapitre de Beaujeu." *Gutenberg-Jahrbuch* 13: 9-16.

742   ———. 1960. "La Papeterie du Suchet aux Ardillats." *Papiergeschichte* 10 (May): 22-28.

743   ———. 1936. "Vieux Moulins a Papier du Beaujolais." *Contribution à l'Histoire de la Papeterie en France* 4: 1-98.

744   ———, and André Blum. 1943. "Le Centre Papetier Ambert-Beaujeu-Annonay." *Contribution à l'Histoire de la Papeterie en France* 9: 19-75.

745   Babler, Otto F. 1934. "Papiermacherei und Aberglaube." *Zeitschrift für Büchfreunde (Third Series)* 3 (May): 111.

746   Blanchet, Augustin. 1910. "Le Papier et Sa Fabrication à Travers les Âges. Découverte de l'Art de Faire le Papier. Origine et Développement de l'Industrie Papetière en France." *Le Bibliofilia* 12 (May): 45-66.

747   Blum, André. 1943. L'Invention du Papier et de l'Imprimerie." *Contribution à l'Histoire de la Papeterie en France* 9: 7-18.

748   ———. 1935. *Les Origines du Papier de l'Imprimerie et de la Gravure.* Paris, France: Éditions de la Tournelle.

749   Bockwitz, Hans H. 1950. "Dokumente Zur Papiergeschichte: ein Französischer Forschungsreisender des 18. Jahrhunderts Über die Beschreibstoffe der Inder." *Wochenblatt für Papierfabrikation* 78 (May 15): 251.

750   ———. 1939. "Zur Lebensgeschichte des Erfinders der Papiermaschine." *Archiv für Buchgewerbe und Gebrauchsgraphik* 76 (August): XXXVI.

751   Bourrachot, Lucile. 1960. "Les Anciennes Papeteries de l'Agenais." *Papiergeschichte* 10 (September): 41-52.

752   ———. 1960. "Les Anciennes Papeteries de l'Agenais." *Papiergeschichte* 10 (November): 53-64.

753   ———. 1961. "Les Anciennes Papeteries de l'Agenais." *Papiergeschichte* 11 (April): 21-31.

754   ———. 1964. "Les Anciennes Papeteries de l'Albret et du Condomois." *Papiergeschichte* 14 (December): 65-72.

755   ———. 1965. "Les Anciennes Papeteries de l'Albret et du Condomois." *Papiergeschichte* 15 (April): 21-28.

756   Briquet, C.M. 1897. "Associations et Grèves des Ouvriers Papetiers

en France aux XVIIe et XVIIIe Siècles." *Revue Internationale de Sociologie* 5 (March): 161-176.

757   Chabrol, G. 1935. "A Propos du Centenaire d'Aristide Bergès." *Contribution à l'Histoire de la Papeterie en France* 2: 87-90.

758   Chobaut, H. 1930. "Les Débuts de l'Industrie du Papier Dans le Comtat-Venaissin — XIVe-XVe Siècles." *Le Bibliographe Moderne* 24: 157-215.

759   ————. 1941. "Le Moulin à Papier de Cadenet (Vaucluse) (1504 Vers 1750)." *Contribution à l'Histoire de la Papeterie en France* 7: 35-42.

760   Churchill, W.A. 1935. *Watermarks in Paper in Holland, England, France, etc , in the XVII and XVIII Centuries and Their Interconnection.* Amsterdam, The Netherlands: M. Hertzberger & Company.

761   Clouzot, H., and C. Follot. 1935. *Histoire du Papier Peint en France.* Paris, France: Charles Moreau.

762   Cohendy, Michel. 1862. "Notes Sur la Papeterie d'Auvergne Antérieurement à 1790 et les Marques de Fabrique des Papeteries de la Ville et Baronnie d'Aubert et Ses Environs." *Mémoires de l'Académie de Science, Belles-Lettres et Arts de Clermont* 4: 196-217.

763   Corraze, Raymond. 1935. "L'Industrie du Papier à Toulouse (1500-1530)." *Contribution à l'Histoire de la Papeterie en France* 2: 95-104.

764   ————. 1941. "Un Moulin à Papier à Toulouse au Commencement du 15e Siècle (1419)." *Contribution à l'Histoire de la Papeterie en France* 6: 13-19.

765   ————. 1941. "Les Papeteries de l'Albigeois et du Castrais aux XVIIe et XVIIIe Siècles." *Contribution à l'Histoire de la Papeterie en France* 6: 71-79.

766   Creveaux, Eugene. 1935. "Les Anciennes Papeteries des Ardennes." *Contribution à l'Histoire de la Papeterie en France* 2: 25-74.

767   ————. 1937. "Un Grand Ingénieur Papetier: Jean-Guillaume Ecrevisse Collaborateur de Nicolas Desarest." *Contribution à l'Histoire de la Papeterie en France* 5: 10-72.

768   ————. 1935. "La Médaille de Reveillon." *Contribution à l'Histoire de la Papeterie en France* 2: 75-78.

769   Deckert, Rudolf. 1933. "Das 'Papier d'Ambert'." *Der Papier-Fabrikant* 31 (July A): 79-80.

770   Degaast, Georges. 1936. "Les Vieux Moulins à Papier d'Auvergne." *Gutenberg-Jahrbuch* 11: 9-13.

771   Deléon, Marcel. 1935. "Le Centenaire d'Aristide Bergès et la Papeterie." *Contribution à l'Histoire de la Papeterie en France* 2: 81-86.

772   Gachet, Henri. 1964. "Botanique & Papeterie. Jean Etienne Guettard 1715-1786." *Papiergeschichte* 14 (April): 23-24.

773   _____. 1964. "Botanique & Papeterie. Jean Etienne Guettard 1715-1786." *Papiergeschichte* 14 (July): 25-28.

774   _____. 1971. "Die Einführung des Holländers in der Französischen Papiermacherei im 18. Jahrhunderts." *Papiergeschichte* 21 (October): 9-15.

775   _____. 1952. "Un Émule Malheureux de N.L. Robert Ferdinand Leistenschneider." *Papiergeschichte* 2 (June): 30-33.

776   _____. 1952. "Un Grand Papyrographe Français Disparait Alexandre Nicolaï." *Papiergeschichte* 2 (November): 61-63.

777   _____. 1971. "Introducción de los Cilindros Holandeses en la Industria Papelera Francesa en el Siglo XVIII." *Investigación y Técnica del Papel* 8 (October): 1067-1079.

778   Gandilhon, René. 1935. "Imprimeurs et Papiers du Midi de la France." *Contribution à l'Histoire de la Papeterie en France* 2: 91-94.

779   Gauthier, Jules. 1897. *L'Industrie du Papier Dans les Hautes Vallées Franc-Comtoises du XVe au XVIIIe Siècles.* Montbeliard, France: Société d'Émulation de Montbéliard.

780   Heitz, Paul. 1902. *Les Filigranes des Papiers Contenus Dans les Archives de la Ville de Strasbourg.* Strasbourg, France: Heitz & Mündel.

781   _____. 1903. *Les Filigranes des Papiers Contenus Dans les Incunables Strasbourgeois de la Bibliothèque Impériale de Strasbourg.* Strasbourg, France: Heitz & Mündel.

782   Janot, Jean-Marie. 1952. *Les Moulins a Papier de la Région Vosgienne.* Nancy, France: Imprimerie Berger-Levrault. 2 Volumes.

783   Lacombe, Henri. 1935. "L'École de Papeterie d'Angoulême." *Contribution à l'Histoire de la Papeterie en France* 2: 131.

784   _____. 1935. "Filigranes de l'Angoumois." *Contribution à l'Histoire de la Papeterie en France* 2: 133-135.

785   _____. 1941. "Le Moulin de Collas." *Contribution à l'Histoire de la Papeterie en France* 7: 19-33.

786   _____. 1935. "Les 'Vins' des Papetiers d'Angoumois." *Contribution à l'Histoire de la Papeterie en France* 2: 125-130.

787   Lebon, Victor. 1935. "Un Grand Ingénieur Papetier: Jehan-Guillaume Ecrevisse." *Contribution à l'Histoire de la Papeterie en France* 2: 137-138.

788   Letonnelier, G. 1941. "Moulin à Papier Sur la Gère." *Contribution à l'Histoire de la Papeterie en France* 7: 43-46.

789   Marchal, Jean. 1962. "Vicissitudes des Anciennes Papeteries Ardennaises." *Papiergeschichte* 12 (December): 41-61.

790   Mateu y Llopis, Felipe. 1967. "Plegado del Papel en la Cancillería de Aragón en el Siglo XIV." *Gutenberg-Jahrbuch* 42: 11-13.

791   Midoux, Etienne, and Auguste Matton. 1868. *Étude Sur les Filigranes des Papiers Employés en France aux XIVe et XVe Siècles.* Paris, France: Dumoulin.

792   Monteil, Casimir, and Jacques Crouau. 1971. *Histoire des Papeteries de Gascogne de la Fondation au 31 Décembre 1970.* Cahors, France: Imprimerie Tardy-Quercy-Auvergne.

793   Morris, Henry. 1976. "More Adventures in Papermaking, etc." In *A Pair on Paper: Two Essays on Paper History and Related Matters,* by Leonard B. Schlosser and Henry Morris. North Hills, Pennsylvania: Bird & Bull Press. 7-43.

794   Nicolaï, Alexandre. 1935. "Considerations Sur l'Origine des Premières Papeteries Françaises." *Contribution à l'Histoire de la Papeterie en France* 2: 117-123.

795   _____. 1935. *Histoire des Moulins à Papier du Sud-Ouest de la France, 1300-1800.* Bordeaux, France: G. Delmas. 2 Volumes.

796   _____. 1941. "Les Moulins à Papier de l'Agenais." *Contribution à l'Histoire de la Papeterie en France* 6: 81-95.

797   _____. 1935. "Note Sur les 'Naypiers' de Toulouse." *Contribution à l'Histoire de la Papeterie en France* 2: 105-107.

798   _____. 1936. "Le Symbolisme Chrétien Dans les Filigranes du Papier." *Contribution à l'Histoire de la Papeterie en France* 3: 7-64.

799   Onfroy, Henri. 1912. *Histoire des Papeteries à la Cuve d'Arches et d'Archettes (1492-1911).* Evreux, France: C. Hérissey. 3rd Edition.

800   Pourrat, Henri. 1936. *Die Alten Papiermühlen in der Auvergne.* Wien, Germany: Herbert Reichner Verlag.

801   Quiblier, Léon. 1941. "L'Imprimerie et la Papeterie de Thonon au XVIe Siècle." *Contribution à l'Histoire de la Papeterie en France* 7: 47-50.

802   Renker, Armin. 1956. "The Auvergne: Birthplace of French Papermaking." *The Paper Maker* 25 (September): 29-33.

803 _____. 1954-1955. "Die Auvergne, Mutterboden der Französischen Papiermacherei." *Imprimatur* 12: 145-148.

804 _____. 1953. "Die Erfindung der Papiermaschine." *Gutenberg-Jahrbuch* 28: 30-35.

805 _____. 1958. "A Paper Mill One Hundred Years Ago." *The Paper Maker* 27 (February): 41-45.

806 Rogerie, H. Bourde de la. 1943. "Les Papeteries de la Région de Morlaix Depuis le XVIe Siècle Jusqu'au Commencement du XIXe Siècle." *Contribution à l'Histoire de la Papeterie en France* 8: 7-61.

807 Sarnecki, Kazimierz. 1956. "Wycíeczka w Czternaste Stulecie." *Przegląd Papierniczy* 12 (October): 307-313.

808 Stein, Henri. 1904. "La Papeterie de Saint-Cloud (Près de Paris) au XIVe Siècle." *Le Bibliographe Moderne* 8: 105-112.

809 _____. 1894. "La Papeterie d'Essonnes." *Annales de la Société Historique et Archéologique du Gâtinais* 12: 335-364.

810 Valls i Subirà, Oriol. 1971. "Estudio Sobre la Trituración de los Traspos (1)." *Investigación y Técnica del Papel* 8 (April): 427-448.

811 Voorn, Henk. 1954. "Capita Selecta: From French Papermaking History." *The Paper Maker* 23 (September): 35-43.

812 _____. 1949. "De Familie Didot." *De Papierwereld* 4 (Special Number): 35-37.

813 _____. 1964. "Gaspard Maillol, Papermaker." *The Paper Maker* 33 (March): 35-42.

814 _____. 1958. "Honoré De Balzac and the Papermaking Industry." *The Paper Maker* 27 (September): 13-16.

815 _____. 1949. "Nicolas-Louis Robert." *De Papierwereld* 4 (Special Number): 11-14.

816 Wiener, Lucien. 1893. *Étude Sur les Filigranes des Papiers Lorrains.* Nancy, France: R. Wiener.

GERMANY

817 Angermann, Gertrud. 1961. "Beiträge Zur Geschichte der Papiermacherfamilie Clasing in Hemeringen an der Weser." *Papiergeschichte* 11 (October): 51-59.

818 _____. 1962. "Beiträge Zur Geschichte der Papiermacherfamilie Clasing in Hemeringen an der Weser." *Papiergeschichte* 12 (June): 31-39.

819 Antz, E.L. 1922. "Zur Geschichte der Papiermühlen und der Druckereien." *Der Papier-Fabrikant* 20 (April 16): 478-480.

820 _____. 1932. "Zur Geschichte der Pfälzischen Papiermühlen." *Pfälzisches Museum* 49: 52-54.

821 _____. 1923. "Die Papiermühlen im Gebiet der Kurpfalz und der Heutigen Rheinpfalz." *Mannheimer Geschichtsblatter* 24: 86-91.

822 Bätzing, Gerhard. 1965. "Die Papiermacherfamilie Scheurmann (Schürman) auf den Beiden Papiermühlen Bei Wolfhagen." *Genealogisches Jahrbuch* 5: 77-120.

823 Belani, E. 1933. "Der Anteil Würtembergs an der Papiermacherei in Kärnten." *Wochenblatt für Papierfabrikation* 64 (Special Edition): 16-17.

824 _____. 1941. "Papiermühle Rappin." *Wochenblatt für Papierfabrikation* 72 (January 18): 44.

825 Bockwitz, Hans H. 1939. "Ahasveri Fritschens 'Abhandlung von Denen Papiermachern'." *Wochenblatt für Papierfabrikation* 70 (May 6): 393-394.

826 _____. 1942. "Das Älteste Bildnis Eines Deutschen Papiermachers." *Archiv für Buchgewerbe und Gebrauchsgraphik* 79: 450.

827 _____. 1937. "Das 'Deutsche Geschin' in Heinrich Zeisings 'Theatrum Machinarum'." *Wochenblatt für Papierfabrikation* 68 (October 2): 756.

828 _____. 1937. "Die Deutsche Papiergeschichtsforschung." *Wochenblatt für Papierfabrikation* 68 (November 27): 928-931.

829 _____. 1941. "Die Deutsche Papiergeschichtsforschung und die Forschungsstelle für Papiergeschichte." *Buch und Schrift* 4: 110-116.

830 _____. 1948. "Dokumente der Papiergeschichte." *Wochenblatt für Papierfabrikation* 76 (June): 142.

831 _____. 1949. "Dokumente Zur Papiergeschichte." *Wochenblatt für Papierfabrikation* 77 (August): 268-269.

832 _____. 1937. "Die Frühesten Abbildungen der 'Deutschen Geschirre'." *Wochenblatt für Papierfabrikation* 68 (September 11): 699-700.

833 _____. 1944. "Zur Frühgeschichte der Elsässer Papiermacherei: die Heifmannsche Papiermühle zu Strasburg i. E. und Ihre Erwähnung in Einer ein Halbes Jahrhundert Lang Verschollenen

Strasburger Urkunde von 1441." *Deutschen Buchgewerbe* 2: 112-116.

834 _____. 1948. "Die Frühzeit des Mehrfarbenhochdruckes Unter Verivendung von Weissen und Farbigen Papieren." *Das Papier* 2 (November): 414-415.

835 _____. 1938. "Ein Gedenkjahr der Papiermacherei im Jubeljahr der Buchdruckerkunst." *Der Altenburger Papierer* 12: 748.

836 _____. 1940. "Geschichte der Papiererzeugung im Donauraum." *Wochenblatt für Papierfabrikation* 71 (August 24): 425-426.

837 _____. 1939. "Goethe und Lichtenberg Als Papierfreunde." *Wochenblatt für Papierfabrikation* 70 (February 25): 173-175.

838 _____. 1940. "Die Gutenbergbibel Als Pergament- und Als Papierdruck." *Wochenblatt für Papierfabrikation* 71 (June 22): 307-310.

839 _____. 1947. "Herr Möller aus Leipzig zu Besuch Bei Pastor Schaeffer in Regensburg 1786." *Wochenblatt für Papierfabrikation* 75 (December): 150.

840 _____. 1943. "Lehrlings- und Gesellenwesen im Alten Papiermacherhandwerk." *Archiv für Buchgewerbe und Gebrauchsgraphik* 80: 118-125.

841 _____. 1938. "Die Leipziger Windpapiermühle vom Jahre 1801." *Wochenblatt für Papierfabrikation* 69 (March 5): 205-208.

842 _____. 1937. "Liegt in den Alten Wasserzeichen ein Verborgener Sinn?" *Wochenblatt für Papierfabrikation* 68 (August 14): 618-620.

843 _____. 1938. "Eine Papiergeschichtliche Rarität." *Philobiblon* 10 (May): 199.

844 _____. 1944. "Die Papiermühle Poprad (Deutschendorf) in der Zips." *Deutschen Buchgewerbe* 2: 81.

845 _____. 1936. "Die Papiermühle zu Serrières Bei Neuchâtel und Ihr Privileg vom Jahre 1477." *Archiv für Buchgewerbe und Gebrauchsgraphik* 73: 229-231.

846 _____. 1947. "Uhland Als Papiermacher." *Wochenblatt für Papierfabrikation* 75 (September-October): 111.

847 _____. 1949. "Ulman Stromer und Seine 'Gleismühle'." *Druckspiegel* 4: 479.

848 _____. 1940. "Viereinhalb Jahrhunderte Papiermühle Gegenbach in Baden." *Archiv für Buchgewerbe und Gebrauchsgraphik* 77 (December): 409-410.

849   _____. 1938. "Zur Wirtschaftslage der Papiermacher und Buchdrucker im Zeitalter Gutenbergs." *Wochenblatt für Papierfabrikation* 69 (September 10): 760-763.

850   Braun, F. 1924. "Zur Geschichte der Papierfabrikation und des Papierhandels in Nürnberg." *Wochenblatt für Papierfabrikation* 55 (April 19): 929-932.

851   _____. 1924. "Zur Geschichte der Papierfabrikation und des Papierhandels in Nürnberg." *Wochenblatt für Papierfqbrikation* 55 (May 10): 1140-1142.

852   _____. 1924. "Zur Geschichte der Papierfabrikation und des Papierhandels in Nürnberg." *Wochenblatt für Papierfabrikation* 55 (May 17): 1209-1211.

853   _____. 1924. "Zur Geschichte der Papierfabrikation und des Papierhandels in Nürnberg." *Wochenblatt für Papierfabrikation* 55 (May 24): 1281-1284.

854   Buchmann, Gerhard. 1952. "Die Papiermacher im Leubenguinde Bei Kahla." *Papiergeschichte* 2 (September): 45-50.

855   _____. 1941. "Die Remdaer Universitäts-Papiermühle." *Wochenblatt für Papierfabrikation* 72 (January 11): 15-19.

856   Claas, Wilhelm. 1952. "Von Alten Papiermühlen Konstruktion, Lage und Aufbau in Gelände." *Papiergeschichte* 2 (April): 16-21.

857   _____. 1957. "Hagen Als Vorort der Alten Märkischen Papiermacherei." *Papiergeschichte* 7 (February): 1-12.

858   _____. 1957. "Hagen Als Vorort der Alten Märkischen Papiermacherei." *Papiergeschichte* 7 (April): 18-23.

859   _____. 1957. "Hagen Als Vorort der Alten Märkischen Papiermacherei." *Papiergeschichte* 7 (May): 37-42.

860   _____. 1958. "Die Papiermühle in Wechte Bei Haus Mark in der Früheren Grafschaft Tecklenburg." *Papiergeschichte* 8 (February): 8-15.

861   _____. 1955. "Von Windpapiermühlen der Vergangenhert." *Papiergeschichte* 5 (February): 1-15.

862   _____. 1956. "Wunderwerke Alter Deutscher Technik." *Papiergeschichte* 6 (February): 1-12.

863   Cornely, Berthold. 1944-1949. "Von Aldus Manutius Bis Johann Adolph Engels: ein Entwicklungsgeschichtlicher Beitrag Zum 'Schönen und Färben' des Papier." *Gutenberg-Jahrbuch* 19-24: 35-38.

864   _____. 1942. "Geschichte des Bläuens und Weisstönens von Papier." *Wochenblatt für Papierfabrikation* 73 (May 9): 132-138

865   Dangel, Albert. 1958. "Von Einer Bisher Unbekannten Papiermühle zu Schwäbisch Gmünd." *Papiergeschichte* 8 (November): 61-62.

866   Dörrer, Anton. 1952. "Papiermühlen im Alten Tirol." *Gutenberg-Jahrbuch* 27: 30-33.

867   Ebenhöch, Hermann. 1950. "Geschichte der Papiermühle in Niedergösgen." *Jahrbuch für Solothurnische Geschichte* 23: 115-142.

868   Eberlein, A. 1957-1958. "Papier, Papiermacher, Papiermühlen in Mecklenburg." *Wissenschaftliche Zeitschrift der Universität Rostock* 7: 133-147.

869   Erhardt, H.L. 1952. "Die Schäffers'schen Versuche - Heute Gesehen." *Schweizerisches Gutenbergmuseum* 38: 14-18.

870   Ernst, Gertrud. 1964. "Aus der Geschichte der Papiermühle Michaelsstein Zum Kloster Michaelsstein, Amt-Blankenburg Gehörig und an Dessen Ringmauern Gelegen." *Papiergeschichte* 14 (July): 42-48.

871   ⸺. 1960. "Versippung der Papiermacherfamilie Schmidt aus Kalldorf (Lippe)." *Papiergeschichte* 10 (July): 39-40.

872   Faisst, R. 1916. "Die Papierfabrikation in Höfen Bei Schopfheim und das Lumpensammeln." *Blätter aus der Markgrafschaft* 2: 55-120.

873   Feldhaus, Franz Maria. 1951. "Altholländische Papier-Patente." *Wochenblatt für Papierfabrikation* 79 (February 28): 98.

874   Flodr, Miroslav. 1971. "Literarische Quellen Zur Mühlengeschichte und Filigranologie von Böhmen Mähren und den Schlesischen Randgebieten." *Papiergeschichte* 21 (December): 57-72.

875   ⸺. 1972. "Literarische Quellen Zur Mühlengeschichte und Filigranologie von Böhmen Mähren und den Schlesischen Randgebieten." *Papiergeschichte* 22 (June): 1-12.

876   Gayoso Carreira, Gonzalo. 1966. "Dos Españoles los Gallegos Antonio y Miguel Introdujeron la Fabricación del Papel en Germania y Basilea." *Investigación y Técnica del Papel* 3 (July): 589-611.

877   Gerardy, Theodor. 1959. "Papiergeschichtliches aus den Spruchakten der Göttinger Jurislenfakultat." *Papiergeschichte* 9 (April): 21-24.

878   ⸺. 1963. "Die Papiermühle Arensburg und Ihre Nesselblatt-Wasserzeichen (1604-1650)." *Papiergeschichte* 13 (October): 25-30.

879 ————. 1960. "Die Wasserzeichen des Mit Gutenbergs Kleiner Psaltertype Gedruckten Missale Speciale." *Papiergeschichte* 10 (May): 13-22.

880 Geuenich, Josef. 1959. *Geschichte der Papierindustrie im Düren-Jülicher Wirtschaftsraum.* Düren, Germany: Dürener Druckerei und Verlag Carl Hamel.

881 ————. 1955. "Zur Geschichte der Windpapiermühle in Schophoven (Kreis Düren)." *Papiergeschichte* 5 (February): 15-16.

882 Girkes, J. 1921. "Die Dürener Papierindustrie Ihre Entwicklung und Wirtschafliche Bedeutung." Unpublished Ph.D. dissertation, University of Bonn.

883 Gleisberg, Hermann. 1970. "Ein Edikt Friedrichs des Grossen zu Gunsten der Papiermühlen (1764)." *Papiergeschichte* 20 (June): 9-11.

884 ————. 1969. "Von der Ehemaligen Windpapiermühle zu Stötteritz Bei Leipzig (1801)." *Papiergeschichte* 19 (November): 30-36.

885 ————. 1969. "Vom Mörser Zur Stampfmühle. Die Entwicklung der Technik des Zerstossens." *Papiergeschichte* 19 (June): 16-23.

886 Grempe, P. Max. 1934. "Der Papierer Nach Abraham a Sancta Clara." *Wochenblatt für Papierfabrikation* 65 (April 14): 255.

887 Grosse-Stoltenberg, Robert. 1961. "Die Papiermühle in Hitzkirchen." *Papiergeschichte* 11 (February): 1-3.

888 ————. 1958. "Ein Wirtschaftsbericht Über die Papierfabrikation vor 136 Jahren." *Papiergeschichte* 8 (November): 62-66.

889 Grossmann, Karl. 1967. "Die Deutsche Papierindustrie auf der II. Weltausstellung in Paris im Jahre 1855." *Papiergeschichte* 17 (June): 47-48.

890 ————. 1952. "Geschichte der Papierfabrikation zu Vlotho (1571-1950)." *Papiergeschichte* 2 (September): 51-55.

891 ————. 1953. "Geschichte der Papierfabrikation zu Vlotho (1571-1950)." *Papiergeschichte* 3 (November): 69-74.

892 ————. 1954. "Die Gesellenfahrten des Papiermeisters Georg Bernhard Hanweg aus Vlotho. (1819-1821)." *Papiergeschichte* 4 (September): 54-56.

893 ————. 1958. "Die Gründung der Papiermühle Westigerbach im Jahre 1705." *Papiergeschichte* 8 (December): 77-79.

894 Haemmerle, Albert. 1966. "Die Wasserzeichen des Alois Dessaur (1763-1850)." *Papiergeschichte* 16 (October): 2-4.

895   Hahn, Wilhelm. 1961. "Zur Geschichte der Flensburger Papiermühle." *Papiergeschichte* 11 (May): 33-35.

896   ————. 1967. "Geschichte der Papiermühle im Gute Rantzau." *Papiergeschichte* 17 (June): 43-45.

897   ————. 1969. "Geschichte der Steinfurter Papiermühle." *Papiergeschichte* 19 (November): 45-47.

898   ————. 1957. "Die Papiermacher- und Tuchfabrikantengeschlecht der Günther. Ein Beitrag Zur Geschichte der Rastorfer Papiermühle." *Die Heimat (Neumünster)* 64: 299-301.

899   ————. 1964. "Die Papiermühle Rastorf in Schleswig-Holstein." *Papiergeschichte* 14 (December): 57-60.

900   Hamann, W. 1958. "Die Papiermühle im Hellthal, Gut Testorf." *Die Heimat (Neumünster)* 65: 278-380.

901   Herzberg, Wilhelm. 1949. *Die Schäfferschen Papierversuche.* Hazen-Kabel/Westfalen, Germany: Papierfabrik Kabel AG.

902   Hössle, Friedrich von. 1921. "Alte Papiermühlen der Deutschen Küstenländer." *Der Papier-Fabrikant* 19 (December 23): 1501-1505.

903   ————. 1921. "Alte Papiermühlen der Deutschen Küstenländer." *Der Papier-Fabrikant* 19 (December 30): 1537-1542.

904   ————. 1922. "Alte Papiermühlen der Deutschen Küstenländer." *Der Papier-Fabrikant* 20 (February 5): 131-135.

905   ————. 1922. "Alte Papiermühlen der Deutschen Küstenländer." *Der Papier-Fabrikant* 20 (February 12): 161-164.

906   ————. 1922. "Alte Papiermühlen der Deutschen Küstenländer." *Der Papier-Fabrikant* 20 (February 26): 225-229.

907   ————. 1922. "Alte Papiermühlen der Deutschen Küstenländer." *Der Papier-Fabrikant* 20 (April 2): 401-405.

908   ————. 1922. "Alte Papiermühlen der Deutschen Küstenländer." *Der Papier-Fabrikant* 20 (October 15): 1433-1438.

909   ————. 1922. "Alte Papiermühlen der Deutschen Küstenländer." *Der Papier-Fabrikant* 20 (October 22): 1461-1467.

910   ————. 1922. "Alte Papiermühlen der Deutschen Küstenländer." *Der Papier-Fabrikant* 20 (December 10): 1688-1692.

911   ————. 1922. "Alte Papiermühlen der Deutschen Küstenländer." *Der Papier-Fabrikant* 20 (December 31): 1779-1782.

912   ————. 1923. "Alte Papiermühlen der Deutschen Küstenländer." *Der Papier-Fabrikant* 21 (January 7): 7-10.

913   ————. 1923. "Alte Papiermühlen der Deutschen Küstenländer." *Der Papier-Fabrikant* 21 (February 4): 65-67.

914 ————. 1923. "Alte Papiermühlen der Deutschen Küstenländer."
Der Papier-Fabrikant 21 (April 15): 205-207.

915 ————. 1923. "Alte Papiermühlen der Deutschen Küstenländer."
Der Papier-Fabrikant 21 (June 9): 39-51.

916 ————. 1923. "Alte Papiermühlen der Deutschen Küstenländer."
Der Papier-Fabrikant 21 (October 7): 447-449.

917 ————. 1923. "Alte Papiermühlen der Deutschen Küstenländer."
Der Papier-Fabrikant 21 (November 11): 485-488.

918 ————. 1923. "Alte Papiermühlen der Deutschen Küstenländer."
Der Papier-Fabrikant 21 (November 25): 499-500.

919 ————. 1923. "Alte Papiermühlen der Deutschen Küstenländer."
Der Papier-Fabrikant 21 (December 9): 514-518.

920 ————. 1923. "Alte Papiermühlen der Deutschen Küstenländer."
Der Papier-Fabrikant 21 (December 23): 531-535.

921 ————. 1923. "Alte Papiermühlen der Deutschen Küstenländer."
Der Papier-Fabrikant 21 (December 30): 544-546.

922 ————. 1924. "Alte Papiermühlen der Deutschen Küstenländer."
Der Papier-Fabrikant 22 (March 2): 80-81.

923 ————. 1932. "Alte Papiermühlen der Herzogtümer
Braunschweig-Wolfenbüttel und Braunschweig-Lüneburg
Insgesamt." *Der Papier-Fabrikant* 30 (October 2-9): 589-594.

924 ————. 1932. "Alte Papiermühlen der Herzogtümer
Braunschweig-Wolfenbüttel und Braunschweig-Lüneburg
Insgesamt." *Der Papier-Fabrikant* 30 (October 16): 601-603.

925 ————. 1928. "Alte Papiermühlen der Hessischen Länder." *Der
Papier-Fabrikant* 26 (May 13): 311-314.

926 ————. 1928. "Alte Papiermühlen der Hessischen Länder." *Der
Papier-Fabrikant* 26 (May 27): 337-341.

927 ————. 1928. "Alte Papiermühlen der Hessischen Länder." *Der
Papier-Fabrikant* 26 (June A): 47-58.

928 ————. 1928. "Alte Papiermühlen der Hessischen Länder." *Der
Papier-Fabrikant* 26 (June 3): 351-358.

929 ————. 1928. "Alte Papiermühlen der Hessischen Länder." *Der
Papier-Fabrikant* 26 (October 28): 681-684.

930 ————. 1928. "Alte Papiermühlen der Hessischen Länder." *Der
Papier-Fabrikant* 26 (November 11): 712-716.

931 ————. 1928. "Alte Papiermühlen der Hessischen Länder." *Der
Papier-Fabrikant* 26 (December 9): 778-783.

932 ————. 1928. "Alte Papiermühlen der Hessischen Länder." *Der
Papier-Fabrikant* 26 (December 16): 793-798.

933 _____. 1929. "Alte Papiermühlen der Hessischen Länder." *Der Papier-Fabrikant* 27 (March 17): 166-170.

934 _____. 1929. "Alte Papiermühlen der Hessischen Länder." *Der Papier-Fabrikant* 27 (April 14): 233-237.

935 _____. 1929. "Alte Papiermühlen der Hessischen Länder." *Der Papier-Fabrikant* 27 (October 20): 647-650.

936 _____. 1929. "Alte Papiermühlen der Hessischen Länder." *Der Papier-Fabrikant* 27 (November 3): 682-690.

937 _____. 1929. "Alte Papiermühlen der Hessischen Länder." *Der Papier-Fabrikant* 27 (November 10): 695-699.

938 _____. 1929. "Alte Papiermühlen der Hessischen Länder." *Der Papier-Fabrikant* 27 (November 24): 728-730.

939 _____. 1934. "Alte Papiermühlen der Preussischen Provinzen West- und Ostpreussen Mit Danzig, Auch Posen." *Der Papier-Fabrikant* 32 (February 25): 90-91.

940 _____. 1934. "Alte Papiermühlen der Preussischen Provinzen West- und Ostpreussen Mit Danzig, Auch Posen." *Der Papier-Fabrikant* 32 (March 4): 99-102.

941 _____. 1934. "Alte Papiermühlen der Preussischen Provinzen West- und Ostpreussen Mit Danzig, Auch Posen." *Der Papier-Fabrikant* 32 (April 29): 193-196.

942 _____. 1934. "Alte Papiermühlen der Preussischen Provinzen West- und Ostpreussen Mit Danzig, Auch Posen." *Der Papier-Fabrikant* 32 (May 6): 208-212.

943 _____. 1934. "Alte Papiermühlen der Preussischen Provinzen West- und Ostpreussen Mit Danzig, Auch Posen." *Der Papier-Fabrikant* 32 (May 20): 231-234.

944 _____. 1933. "Alte Papiermühlen der Provinz Brandenburg." *Der Papier-Fabrikant* 31 (January 1): 1-4.

945 _____. 1933. "Alte Papiermühlen der Provinz Brandenburg." *Der Papier-Fabrikant* 31 (January 8): 15-20.

946 _____. 1933. "Alte Papiermühlen der Provinz Brandenburg." *Der Papier-Fabrikant* 31 (April 9): 233-236.

947 _____. 1933. "Alte Papiermühlen der Provinz Brandenburg." *Der Papier-Fabrikant* 31 (April 16): 248-251.

948 _____. 1933. "Alte Papiermühlen der Provinz Brandenburg." *Der Papier-Fabrikant* 31 (April 23): 260-261.

949 _____. 1933. "Alte Papiermühlen der Provinz Brandenburg." *Der Papier-Fabrikant* 31 (June 18): 353-354.

950 _____. 1933. "Alte Papiermühlen der Provinz Brandenburg." *Der Papier-Fabrikant* 31 (June 25): 365-369.

951 ———. 1933. "Alte Papiermühlen der Provinz Brandenburg." *Der Papier-Fabrikant* 31 (September 10): 489-490.

952 ———. 1933. "Alte Papiermühlen der Provinz Brandenburg." *Der Papier-Fabrikant* 31 (September 17): 500-506.

953 ———. 1933. "Alte Papiermühlen der Provinz Brandenburg." *Der Papier-Fabrikant* 31 (September 24): 510-516.

954 ———. 1933. "Alte Papiermühlen der Provinz Brandenburg." *Der Papier-Fabrikant* 31 (October 1): 524-526.

955 ———. 1933. "Alte Papiermühlen der Provinz Brandenburg." *Der Papier-Fabrikant* 31 (November 26): 630-634.

956 ———. 1933. "Alte Papiermühlen der Provinz Brandenburg." *Der Papier-Fabrikant* 31 (December 3): 643-648.

957 ———. 1929. "Alte Papiermühlen der Provinz Sachsen." *Wochenblatt für Papierfabrikation* 60 (July 13): 856-861.

958 ———. 1929. "Alte Papiermühlen der Provinz Sachsen." *Wochenblatt für Papierfabrikation* 60 (August 10): 982-984.

959 ———. 1929. "Alte Papiermühlen der Provinz Sachsen." *Wochenblatt für Papierfabrikation* 60 (September 7): 1103-1107.

960 ———. 1929. "Alte Papiermühlen der Provinz Sachsen." *Wochenblatt für Papierfabrikation* 60 (October 26): 1330-1333.

961 ———. 1929. "Alte Papiermühlen der Provinz Sachsen." *Wochenblatt für Papierfabrikation* 60 (December 28): 1636-1641.

962 ———. 1930. "Alte Papiermühlen der Provinz Sachsen." *Wochenblatt für Papierfabrikation* 61 (August 9): 1034-1039.

963 ———. 1930. "Alte Papiermühlen der Provinz Sachsen." *Wochenblatt für Papierfabrikation* 61 (September 20): 1218-1221.

964 ———. 1930. "Alte Papiermühlen der Provinz Sachsen." *Wochenblatt für Papierfabrikation* 61 (December 20): 1649-1651.

965 ———. 1930. "Alte Papiermühlen der Provinz Sachsen." *Wochenblatt für Papierfabrikation* 61 (December 27): 1679-1684.

966 ———. 1931. "Alte Papiermühlen der Provinz Sachsen." *Wochenblatt für Papierfabrikation* 62 (May 2): 423-428.

967 ———. 1931. "Alte Papiermühlen der Provinz Sachsen." *Wochenblatt für Papierfabrikation* 62 (May 30): 519-523.

968 _____. 1931. "Alte Papiermühlen der Provinz Sachsen."
Wochenblatt für Papierfabrikation 62 (June 6A): 13-17.
969 _____. 1931. "Alte Papiermühlen der Provinz Sachsen."
Wochenblatt für Papierfabrikation 62 (August 15): 783-784.
970 _____. 1931. "Alte Papiermühlen der Provinz Sachsen."
Wochenblatt für Papierfabrikation 62 (September 19): 900-902.
971 _____. 1932. "Alte Papiermühlen der Provinz Sachsen."
Wochenblatt für Papierfabrikation 63 (April 2): 260-262.
972 _____. 1932. "Alte Papiermühlen der Provinz Sachsen."
Wochenblatt für Papierfabrikation 63 (July 30): 588-590.
973 _____. 1932. "Alte Papiermühlen der Provinz Sachsen."
Wochenblatt für Papierfabrikation 63 (August 20): 640-641.
974 _____. 1932. "Alte Papiermühlen der Provinz Sachsen."
Wochenblatt für Papierfabrikation 63 (December 24): 951-954.
975 _____. 1933. "Alte Papiermühlen der Provinz Sachsen."
Wochenblatt für Papierfabrikation 64 (April 8): 239-240.
976 _____. 1933. "Alte Papiermühlen der Provinz Sachsen."
Wochenblatt für Papierfabrikation 64 (June 30): 450-451.
977 _____. 1933. "Alte Papiermühlen der Provinz Sachsen."
Wochenblatt für Papierfabrikation 64 (July 15): 491-494.
978 _____. 1933. "Alte Papiermühlen der Provinz Sachsen."
Wochenblatt für Papierfabrikation 64 (November 4): 770-772.
979 _____. 1933. "Alte Papiermühlen der Provinz Sachsen."
Wochenblatt für Papierfabrikation 64 (November 18):
806-807.
980 _____. 1934. "Alte Papiermühlen der Provinz Sachsen."
Wochenblatt für Papierfabrikation 65 (May 19): 356-358.
981 _____. 1934. "Alte Papiermühlen der Provinz Sachsen."
Wochenblatt für Papierfabrikation 65 (July 7): 483-485.
982 _____. 1934. "Alte Papiermühlen der Provinz Sachsen."
Wochenblatt für Papierfabrikation 65 (July 28): 540-542.
983 _____. 1934. "Alte Papiermühlen der Provinz Sachsen."
Wochenblatt für Papierfabrikation 65 (December 8): 872-873.
984 _____. 1935. "Alte Papiermühlen der Provinz Sachsen."
Wochenblatt für Papierfabrikation 66 (January 12): 37-38.
985 _____. 1935. "Alte Papiermühlen der Provinz Sachsen."
Wochenblatt für Papierfabrikation 66 (February 16): 113-116.
986 _____. 1935. "Alte Papiermühlen der Provinz Sachsen."
Wochenblatt für Papierfabrikation 66 (April 27): 329-333.
987 _____. 1935. "Alte Papiermühlen der Provinz Sachsen."
Wochenblatt für Papierfabrikation 66 (July 20): 556-559.

988   _____. 1935. "Alte Papiermühlen der Provinz Sachsen."
Wochenblatt für Papierfabrikation 66 (August 24): 646-647.

989   _____. 1935. "Alte Papiermühlen der Provinz Sachsen."
Wochenblatt für Papierfabrikation 66 (October 5): 756-759.

990   _____. 1935. "Alte Papiermühlen der Provinz Sachsen."
Wochenblatt für Papierfabrikation 66 (December 28):
977-980.

991   _____. 1936. "Alte Papiermühlen der Provinz Sachsen,"
Wochenblatt für Papierfabrikation 67 (April 11): 278-280.

992   _____. 1936. "Alte Papiermühlen der Provinz Sachsen."
Wochenblatt für Papierfabrikation 67 (August 22): 632-634.

993   _____. 1936. "Alte Papiermühlen der Provinz Sachsen."
Wochenblatt für Papierfabrikation 67 (September 12):
689-691.

994   _____. 1936. "Alte Papiermühlen der Provinz Sachsen."
Wochenblatt für Papierfabrikation 67 (October 24): 801-803.

995   _____. 1937. "Alte Papiermühlen der Provinz Sachsen."
Wochenblatt für Papierfabrikation 68 (May 15): 371-373.

996   _____. 1937. "Alte Papiermühlen der Provinz Sachsen."
Wochenblatt für Papierfabrikation 68 (August 21): 641-642.

997   _____. 1937. "Alte Papiermühlen der Provinz Sachsen."
Wochenblatt für Papierfabrikation 68 (September 18):
715-716.

998   _____. 1935. "Alte Papiermühlen der Provinz Schlesien." *Der*
Papier-Fabrikant 33 (January 13): 9-13.

999   _____. 1935. "Alte Papiermühlen der Provinz Schlesien." *Der*
Papier-Fabrikant 33 (January 20): 19-22.

1000   _____. 1935. "Alte Papiermühlen der Provinz Schlesien." *Der*
Papier-Fabrikant 33 (February 3): 38-40.

1001   _____. 1935. "Alte Papiermühlen der Provinz Schlesien." *Der*
Papier-Fabrikant 33 (June 30): 221-223.

1002   _____. 1935. "Alte Papiermühlen der Provinz Schlesien." *Der*
Papier-Fabrikant 33 (July 7): 227-228.

1003   _____. 1935. "Alte Papiermühlen der Provinz Schlesien." *Der*
Papier-Fabrikant 33 (July 14): 236-238.

1004   _____. 1935. "Alte Papiermühlen der Provinz Schlesien." *Der*
Papier-Fabrikant 33 (July 21): 246-248.

1005   _____. 1935. "Alte Papiermühlen der Provinz Schlesien." *Der*
Papier-Fabrikant 33 (August 4): 260-262.

1006   _____. 1935. "Alte Papiermühlen der Provinz Schlesien." *Der*
Papier-Fabrikant 33 (August 11): 269-272.

1007 ————. 1935. "Alte Papiermühlen der Provinz Schlesien." *Der Papier-Fabrikant* 33 (August 18): 279-280.

1008 ————. 1935. "Alte Papiermühlen der Provinz Schlesien." *Der Papier-Fabrikant* 33 (September 1): 293-296.

1009 ————. 1935. "Alte Papiermühlen der Provinz Schlesien." *Der Papier-Fabrikant* 33 (October 27): 357-359.

1010 ————. 1935. "Alte Papiermühlen der Provinz Schlesien." *Der Papier-Fabrikant* 33 (November 10): 369-372.

1011 ————. 1935. "Alte Papiermühlen der Provinz Schlesien." *Der Papier-Fabrikant* 33 (December 8): 427-431.

1012 ————. 1935. "Alte Papiermühlen der Provinz Schlesien." *Der Papier-Fabrikant* 33 (December 27): 446-448.

1013 ————. 1936. "Alte Papiermühlen der Provinz Schlesien." *Der Papier-Fabrikant* 34 (January 12): 14-15.

1014 ————. 1936. "Alte Papiermühlen der Provinz Schlesien." *Der Papier-Fabrikant* 34 (January 26): 29-31.

1015 ————. 1936. "Alte Papiermühlen der Provinz Schlesien." *Der Papier-Fabrikant* 34 (February 2): 37-39.

1016 ————. 1936. "Alte Papiermühlen der Provinz Schlesien." *Der Papier-Fabrikant* 34 (May 17): 155-159.

1017 ————. 1936. "Alte Papiermühlen der Provinz Schlesien." *Der Papier-Fabrikant* 34 (May 24): 163-166.

1018 ————. 1936. "Alte Papiermühlen der Provinz Schlesien." *Der Papier-Fabrikant* 34 (August 9): 302-304.

1019 ————. 1936. "Alte Papiermühlen der Provinz Schlesien." *Der Papier-Fabrikant* 34 (August 16): 310-312.

1020 ————. 1936. "Alte Papiermühlen der Provinz Schlesien." *Der Papier-Fabrikant* 34 (September 6): 333-335.

1021 ————. 1936. "Alte Papiermühlen der Provinz Schlesien." *Der Papier-Fabrikant* 34 (September 13): 340-342.

1022 ————. 1938. "Alte Papiermühlen der Provinz Schlesien." *Der Papier-Fabrikant* 36 (February 25): 73-79.

1023 ————. 1938. "Alte Papiermühlen der Provinz Schlesien." *Der Papier-Fabrikant* 36 (March 11): 93-95.

1024 ————. 1938. "Alte Papiermühlen der Provinz Schlesien." *Der Papier-Fabrikant* 36 (March 18): 101-103.

1025 ————. 1928. "Alte Papiermühlen der Provinz Westfalen." *Wochenblatt für Papierfabrikation* 59 (March 31): 340-345.

1026 ————. 1928. "Alte Papiermühlen der Provinz Westfalen." *Wochenblatt für Papierfabrikation* 59 (May 19): 539-541.

1027 ————. 1928. "Alte Papiermühlen der Provinz Westfalen."
Wochenblatt für Papierfabrikation 59 (August 11): 877-882.

1028 ————. 1928. "Alte Papiermühlen der Provinz Westfalen."
Wochenblatt für Papierfabrikation 59 (October 6): 1111-1114.

1029 ————. 1928. "Alte Papiermühlen der Provinz Westfalen."
Wochenblatt für Papierfabrikation 59 (November 3):
1234-1236.

1030 ————. 1928. "Alte Papiermühlen der Provinz Westfalen."
Wochenblatt für Papierfabrikation 59 (December 1):
1365-1369.

1031 ————. 1928. "Alte Papiermühlen der Provinz Westfalen."
Wochenblatt für Papierfabrikation 59 (December 22):
1475-1476.

1032 ————. 1926. "Alte Papiermühlen der Rheinprovinz."
Wochenblatt für Papierfabrikation 57 (April 24): 473-478.

1033 ————. 1926. "Alte Papiermühlen der Rheinprovinz."
Wochenblatt für Papierfabrikation 57 (May 22): 581-584.

1034 ————. 1926. "Alte Papiermühlen der Rheinprovinz."
Wochenblatt für Papierfabrikation 57 (June 12A): 11-27.

1035 ————. 1926. "Alte Papiermühlen der Rheinprovinz."
Wochenblatt für Papierfabrikation 57 (July 10): 773-776.

1036 ————. 1926. "Alte Papiermühlen der Rheinprovinz."
Wochenblatt für Papierfabrikation 57 (July 31): 849-853.

1037 ————. 1926. "Alte Papiermühlen der Rheinprovinz."
Wochenblatt für Papierfabrikation 57 (October 16):
1149-1151.

1038 ————. 1926. "Alte Papiermühlen der Rheinprovinz."
Wochenblatt für Papierfabrikation 57 (November 6):
1238-1240.

1039 ————. 1926. "Alte Papiermühlen der Rheinprovinz."
Wochenblatt für Papierfabrikation 57 (December 18):
1432-1436.

1040 ————. 1927. "Alte Papiermühlen der Rheinprovinz."
Wochenblatt für Papierfabrikation 58 (February 12): 169-171.

1041 ————. 1927. "Alte Papiermühlen der Rheinprovinz."
Wochenblatt für Papierfabrikation 58 (March 31): 383-386.

1042 ————. 1927. "Alte Papiermühlen der Rheinprovinz."
Wochenblatt für Papierfabrikation 58 (April 30): 503-505.

1043 ————. 1927. "Alte Papiermühlen der Rheinprovinz."
Wochenblatt für Papierfabrikation 58 (May 14): 573-577.

1044 ————. 1927. "Alte Papiermühlen der Rheinprovinz."
Wochenblatt für Papierfabrikation 58 (May 21): 614-617.

1045 ————. 1927. "Alte Papiermühlen der Rheinprovinz."
Wochenblatt für Papierfabrikation 58 (June 18A): 14-25.

1046 ————. 1927. "Alte Papiermühlen der Rheinprovinz."
Wochenblatt für Papierfabrikation 58 (June 18): 745.

1047 ————. 1927. "Alte Papiermühlen der Rheinprovinz."
Wochenblatt für Papierfabrikation 58 (September 3):
1070-1075.

1048 ————. 1927. "Alte Papiermühlen der Rheinprovinz."
Wochenblatt für Papierfabrikation 58 (October 22):
1283-1287.

1049 ————. 1927. "Alte Papiermühlen der Rheinprovinz."
Wochenblatt für Papierfabrikation 58 (November 19):
1403-1408.

1050 ————. 1927. "Alte Papiermühlen der Rheinprovinz."
Wochenblatt für Papierfabrikation 58 (December 17):
1532-1536.

1051 ————. 1927. "Alte Papiermühlen der Rheinprovinz."
Wochenblatt für Papierfabrikation 58 (December 31):
1593-1595.

1052 ————. 1930. "Alte Papiermühlen im Herzogtum Braunschweig."
Der Papier-Fabrikant 28 (May 4): 303-305.

1053 ————. 1930. "Alte Papiermühlen im Herzogtum Braunschweig."
Der Papier-Fabrikant 28 (May 11): 319-320.

1054 ————. 1930. "Alte Papiermühlen im Herzogtum Braunschweig."
Der Papier-Fabrikant 28 (May 25): 347-351.

1055 ————. 1930. "Alte Papiermühlen im Herzogtum Braunschweig."
Der Papier-Fabrikant 28 (June 1): 366-369.

1056 ————. 1930. "Alte Papiermühlen im Herzogtum Braunschweig."
Der Papier-Fabrikant 28 (June 15): 394-397.

1057 ————. 1931. "Alte Papiermühlen im Herzogtum Branschweig-
Lüneburg, dem Späteren Königreich, dann Preussische Provinz
Hannover." *Der Papier-Fabrikant* 29 (March 22): 177-181.

1058 ————. 1931. "Alte Papiermühlen im Herzogtum Braunschweig-
Lüneburg, dem Späteren Königreich, dann Preussische Provinz
Hannover." *Der Papier-Fabrikant* 29 (April 12): 228-231.

1059 ————. 1931. "Alte Papiermühlen im Herzogtum Braunschweig-
Lüneburg, dem Späteren Königreich, dann Preussische Provinz
Hannover." *Der Papier-Fabrikant* 29 (May 3): 280-284.

1060 ————. 1931. "Alte Papiermühlen im Herzogtum Braunschweig-
Lüneburg, dem Späteren Königreich, dann Preussische Provinz
Hannover." *Der Papier-Fabrikant* 29 (October 4): 637-640.

1061 ————. 1931. "Alte Papiermühlen im Herzogtum Braunschweig-
Lüneburg, dem Späteren Königreich, dann Preussische Provinz
Hannover." *Der Papier-Fabrikant* 29 (October 11): 651-658.

1062 ————. 1932. "Alte Papiermühlen im Herzogtum Braunschweig-
Lüneburg, dem Späteren Königreich, dann Preussische Provinz
Hannover." *Der Papier-Fabrikant* 30 (January 3): 5-9.

1063 ————. 1932. "Alte Papiermühlen im Herzogtum Braunschweig-
Lüneburg, dem Späteren Königreich, dann Preussische Provinz
Hannover." *Der Papier-Fabrikant* 30 (February 14): 80.

1064 ————. 1932. "Alte Papiermühlen im Herzogtum Braunschweig-
Lüneburg, dem Späteren Königreich, dann Preussische Provinz
Hannover." *Der Papier-Fabrikant* 30 (March 20): 134-138.

1065 ————. 1932. "Alte Papiermühlen im Herzogtum Braunschweig-
Lüneburg, dem Späteren Königreich, dann Preussische Provinz
Hannover." *Der Papier-Fabrikant* 30 (June 5): 363-369.

1066 ————. 1929. "Alte Papiermühlen in der Rheinprovinz,
Westfalen, Waldeck und Lippe." *Wochenblatt für
Papierfabrikation* 60 (February 16): 195-198.

1067 ————. 1929. "Alte Papiermühlen in der Rheinprovinz,
Westfalen, Waldeck und Lippe." *Wochenblatt für
Papierfabrikation* 60 (February 23): 229-232.

1068 ————. 1929. "Alte Papiermühlen in der Rheinprovinz,
Westfalen, Waldeck und Lippe." *Wochenblatt für
Papierfabrikation* 60 (April 13): 447-450.

1069 ————. 1929. "Alte Papiermühlen in der Rheinprovinz,
Westfalen, Waldeck und Lippe." *Wochenblatt für
Papierfabrikation* 60 (April 27): 509-511.

1070 ————. 1929. "Alte Papiermühlen in der Rheinprovinz,
Westfalen, Waldeck und Lippe." *Wochenblatt für
Papierfabrikation* 60 (May 11): 578-581.

1071 ————. 1921. "Alte Pfälzische Papiermühlen." *Der Papier-
Fabrikant* 19 (January 14): 25-28.

1072 ————. 1921. "Alte Pfälzische Papiermühlen." *Der Papier-
Fabrikant* 19 (February 11): 117-122.

1073 ————. 1921. "Alte Pfälzische Papiermühlen." *Der Papier-
Fabrikant* 19 (June A): 16-27.

1074 ————. 1923. "Alte Pfälzische Papiermühlen." *Der Papier-
Fabrikant* 21 (March 18): 158-160.

1075 —————. 1924. "Die Alten Papiermacher Fischer." *Wochenblatt für Papierfabrikation* 55 (June 14A): 17-30.

1076 —————. 1925. "Die Alten Papiermacher Fischer und die Mustauer Papiermühle." *Wochenblatt für Papierfabrikation* 56 (June 13A): 8-23.

1077 —————. 1907. *Die Alten Papiermühlen der Freien Reichstadt Augsburg Sowie Alte Papiere und Deren Wasserzeichen in Stadt-Archiv und der Kreis- und Stadt-Bibliothek zu Augsburg.* Augsburg, Germany: Math. Rieger'schen Buchh.

1078 —————. 1920. "Die Ältesten Papiermühlen im Kirchenstaat und in den Angrenzenden Provinzen." *Der Papier-Fabrikant* 18 (January 23): 56-59.

1079 —————. 1920. "Die Ältesten Papiermühlen im Kirchenstaat und in den Angrenzenden Provinzen." *Der Papier-Fabrikant* 18 (January 30): 71-78.

1080 —————. 1920. "Die Ältesten Papiermühlen im Kirchenstaat und in den Angrenzenden Provinzen." *Der Papier-Fabrikant* 18 (February 13): 121-124.

1081 —————. 1920. "Die Ältesten Papiermühlen im Kirchenstaat und in den Angrenzenden Provinzen." *Der Papier-Fabrikant* 18 (February 20): 143-146.

1082 —————. 1913. "Altschlesische Papiermühlen." *Der Papier-Fabrikant* 11 (June 13A): 31-42.

1083 —————. 1924. "Bayerische Papiergeschichte." *Der Papier-Fabrikant* 22 (March 9): 95-97.

1084 —————. 1924. "Bayerische Papiergeschichte." *Der Papier-Fabrikant* 22 (March 16): 107-109.

1085 —————. 1924. "Bayerische Papiergeschichte." *Der Papier-Fabrikant* 22 (March 30): 136-138.

1086 —————. 1924. "Bayerische Papiergeschichte." *Der Papier-Fabrikant* 22 (April 20): 175-177.

1087 —————. 1924. "Bayerische Papiergeschichte." *Der Papier-Fabrikant* 22 (May 4): 197-200.

1088 —————. 1924. "Bayerische Papiergeschichte." *Der Papier-Fabrikant* 22 (May 25): 232-238.

1089 —————. 1924. "Bayerische Papiergeschichte." *Der Papier-Fabrikant* 22 (June 8): 254-259.

1090 —————. 1924. "Bayerische Papiergeschichte." *Der Papier-Fabrikant* 22 (June 29): 293-296.

1091 —————. 1924. "Bayerische Papiergeschichte." *Der Papier-Fabrikant* 22 (July 13): 319-321.

1092   ———. 1924. "Bayerische Papiergeschichte." *Der Papier-Fabrikant* 22 (July 20): 329-333.

1093   ———. 1924. "Bayerische Papiergeschichte." *Der Papier-Fabrikant* 22 (July 27): 342-344.

1094   ———. 1924. "Bayerische Papiergeschichte." *Der Papier-Fabrikant* 22 (August 24): 388-393.

1095   ———. 1924. "Bayerische Papiergeschichte." *Der Papier-Fabrikant* 22 (September 14): 427-429.

1096   ———. 1924. "Bayerische Papiergeschichte." *Der Papier-Fabrikant* 22 (September 28): 453-456.

1097   ———. 1924. "Bayerische Papiergeschichte." *Der Papier-Fabrikant* 22 (October 19): 496-499.

1098   ———. 1924. "Bayerische Papiergeschichte." *Der Papier-Fabrikant* 22 (November 9): 534-538.

1099   ———. 1924. "Bayerische Papiergeschichte." *Der Papier-Fabrikant* 22 (December 28): 618-620.

1100   ———. 1925. "Bayerische Papiergeschichte." *Der Papier-Fabrikant* 23 (January 25): 47-50.

1101   ———. 1925. "Bayerische Papiergeschichte." *Der Papier-Fabrikant* 23 (February 8): 75-79.

1102   ———. 1925, "Bayerische Papiergeschichte." *Der Papier-Fabrikant* 23 (March 22): 188-191.

1103   ———. 1925. "Bayerische Papiergeschichte." *Der Papier-Fabrikant* 23 (April 19): 262-266.

1104   ———. 1925. "Bayerische Papiergeschichte." *Der Papier-Fabrikant* 23 (May 17): 321-325.

1105   ———. 1925. "Bayerische Papiergeschichte." *Der Papier-Fabrikant* 23 (June 7): 370-373.

1106   ———. 1925. "Bayerische Papiergeschichte." *Der Papier-Fabrikant* 23 (June 14): 388-390.

1107   ———. 1925. "Bayerische Papiergeschichte." *Der Papier-Fabrikant* 23 (June 18): 117-128.

1108   ———. 1925. "Bayerische Papiergeschichte." *Der Papier-Fabrikant* 23 (October 11): 652-655.

1109   ———. 1925. "Bayerische Papiergeschichte." *Der Papier-Fabrikant* 23 (October 18): 671-674.

1110   ———. 1925. "Bayerische Papiergeschichte." *Der Papier-Fabrikant* 23 (October 25): 683-687.

1111   ———. 1925. "Bayerische Papiergeschichte." *Der Papier-Fabrikant* 23 (December 21): 816-824.

1112 _____. 1926. "Bayerische Papiergeschichte." *Der Papier-Fabrikant* 24 (February 21): 117-119.

1113 _____. 1926. "Bayerische Papiergeschichte." *Der Papier-Fabrikant* 24 (February 28): 126-128.

1114 _____. 1926. "Bayerische Papiergeschichte." *Der Papier-Fabrikant* 24 (March 14): 155-157.

1115 _____. 1926. "Bayerische Papiergeschichte." *Der Papier-Fabrikant* 24 (March 28): 189-191.

1116 _____. 1926. "Bayerische Papiergeschichte." *Der Papier-Fabrikant* 24 (April 4): 209-212.

1117 _____. 1926. "Bayerische Papiergeschichte." *Der Papier-Fabrikant* 24 (April 11): 222-226.

1118 _____. 1926. "Bayerische Papiergeschichte." *Der Papier-Fabrikant* 24 (April 18): 238-242.

1119 _____. 1926. "Bayerische Papiergeschichte." *Der Papier-Fabrikant* 24 (May 23): 313-317.

1120 _____. 1926. "Bayerische Papiergeschichte." *Der Papier-Fabrikant* 24 (June 5): 49-64.

1121 _____. 1927. "Bayerische Papiergeschichte." *Der Papier-Fabrikant* 25 (February 13): 102-107.

1122 _____. 1927. "Bayerische Papiergeschichte." *Der Papier-Fabrikant* 25 (March 13): 164-166.

1123 _____. 1927. "Bayerische Papiergeschichte." *Der Papier-Fabrikant* 25 (April 17): 243-247.

1124 _____. 1927. "Bayerische Papiergeschichte." *Der Papier-Fabrikant* 25 (May 15): 302-305.

1125 _____. 1927. "Bayerische Papiergeschichte." *Der Papier-Fabrikant* 25 (June 26): 402-405.

1126 _____. 1927. "Bayerische Papiergeschichte." *Der Papier-Fabrikant* 25 (July 17): 450-452.

1127 _____. 1927. "Bayerische Papiergeschichte." *Der Papier-Fabrikant* 25 (July 31): 485-489.

1128 _____. 1927. "Bayerische Papiergeschichte." *Der Papier-Fabrikant* 25 (August 7): 501-503.

1129 _____. 1927. "Bayerische Papiergeschichte." *Der Papier-Fabrikant* 25 (December 18): 803.

1130 _____. 1920. "Das Christus-Monogramm Als Wasserzeichen. Jesuitische Papiermühlen. Universitätspapiermacher." *Der Papier-Fabrikant* 18 (September 24): 721-723.

1131 _____. 1920. "Das Christus-Monogramm Als Wasserzeichen.

Jesuitische Papiermühlen. Universitätspapiermacher." *Der Papier-Fabrikant* 18 (October 1): 739-742.

1132 _____. 1920. "Das Christus-Monogramm Als Wasserzeichen. Jesuitische Papiermühlen. Universitätspapiermacher." *Der Papier-Fabrikant* 18 (October 8): 762-765.

1133 _____. 1921. "Zur Einführung der Papiermaschine in Deutschland." *Der Papier-Fabrikant* 19 (November 18): 1321-1322.

1134 _____. 1929. "Einige Lumpensammler-Typen." *Der Papier-Fabrikant* 27 (June A): 53-56.

1135 _____. 1926. "Zur Entwicklung der Rheinischen Papierindustrie Einführung der Papiermaschine in Deutschland und der Schweiz." *Der Papier-Fabrikant* 24 (March 7): 148.

1136 _____. 1929. "Ein Gedenkblatt für Ulman Stromer." *Wochenblatt für Papierfabrikation* 60 (January 5): 6-9.

1137 _____. 1900. *Geschichte der Alten Papiermühlen im Ehemaligen Stift Kempten und in der Reichstadt Kempten.* Kempten, Germany: Kösel.

1138 _____. 1921. "Geschichte der Patentpapierfabrik zu Berlin." *Der Papier-Fabrikant* 19 (November 25): 1357-1358.

1139 _____. 1921. *Geschichte des Alten Papiermacherhandwerks im Weyland Heyligen Römischen Reich.* Wien, Germany: Verlag des Zentralblatt für die Papierindustrie.

1140 _____. 1924. "Hans Caspar Escher." *Wochenblatt für Papierfabrikation* 55 (June 14A): 107-109.

1141 _____. 1929. "Hausmarte und Handwertszeichen der Alten Papiermacher und Neuzertliches Papiermachererwappen." *Wochenblatt für Papierfabrikation* 60 (June 8A): 14-19.

1142 _____. 1922. "Ein Lebensbild Ulman Stromers." *Wochenblatt für Papierfabrikation* 53 (June 3A): 20-26.

1143 _____. 1928. "Eine Mecklenburger Papiermühle in 150jahrigen Familienbesitz." *Wochenblatt für Papierfabrikation* 59 (March 24): 322.

1144 _____. 1930. "Nachtrag zu den Hessischen Papiermühlen." *Der Papier-Fabrikant* 28 (May 4): 306-307.

1145 _____. 1930. "Nachträge Zur Bayerischen Papiergeschichte." *Der Papier-Fabrikant* 28 (June A): 58-76.

1146 _____. 1926. "Nochmals Zur Geschichte der Rheinischen Papierindustrie Sowie Einführung der Papiermaschine in Deutschland." *Der Papier-Fabrikant* 24 (July 18): 439-440.

1147   ————. 1922. "Die Papierindustrie in der Bayerischen Industrie-Ausstellung zu Nürnberg im Jahre 1840." *Wochenblatt für Papierfabrikation* 53 (June 3A): 48-50.

1148   ————. 1922. "Die Wasserzeichen und Ihre Bedeutung Als Schutzmarke in Alter und Neuer Zeit." *Zentralblatt für die Papierindustrie* 40: 303-307.

1149   ————. 1919. "Württembergische Papiergeschichte." *Wochenblatt für Papierfabrikation* 50 (August 21): 2024-2027.

1150   ————. 1919. "Württembergische Papiergeschichte." *Wochenblatt für Papierfabrikation* 50 (November 1): 2819-2820.

1151   ————. 1920. "Württembergische Papiergeschichte." *Wochenblatt für Papierfabrikation* 51 (January 10): 25-27.

1152   ————. 1920. "Württembergische Papiergeschichte." *Wochenblatt für Papierfabrikation* 51 (July 24): 2058-2060.

1153   ————. 1920. "Württembergische Papiergeschichte." *Wochenblatt für Papierfabrikation* 51 (August 7): 2183-2184.

1154   ————. 1920. "Württembergische Papiergeschichte." *Wochenblatt für Papierfabrikation* 51 (October 2): 2745-2747.

1155   ————. 1920. "Württembergische Papiergeschichte." *Wochenblatt für Papierfabrikation* 51 (December 4): 3388-3390.

1156   ————. 1921. "Württembergische Papiergeschichte." *Wochenblatt für Papierfabrikation* 52 (January 29): 252-254.

1157   ————. 1921. "Württembergische Papiergeschichte." *Wochenblatt für Papierfabrikation* 52 (February 19): 497-499.

1158   ————. 1921. "Württembergische Papiergeschichte." *Wochenblatt für Papierfabrikation* 52 (May 28): 1617-1626.

1159   ————. 1921. "Württembergische Papiergeschichte." *Wochenblatt für Papierfabrikation* 52 (June 30): 1837-1838.

1160   ————. 1921. "Württembergische Papiergeschichte." *Wochenblatt für Papierfabrikation* 52 (October 15): 3337-3340.

1161   ————. 1921. "Württembergische Papiergeschichte." *Wochenblatt für Papierfabrikation* 52 (December 10): 4044-4048.

1162   ————. 1921. "Württembergische Papiergeschichte." *Wochenblatt für Papierfabrikation* 52 (December 24): 4230-4233.

1163   ————. 1922. "Württembergische Papiergeschichte." *Wochenblatt für Papierfabrikation* 53 (February 4): 394-396.

1164   ————. 1922. "Württembergische Papiergeschichte." *Wochenblatt für Papierfabrikation* 53 (April 29): 1523-1526.

1165   ————. 1923. "Württembergische Papiergeschichte." *Wochenblatt für Papierfabrikation* 54 (June 2): 1506-1515.

1166   ————. 1924. "Württembergische Papiergeschichte." *Wochenblatt für Papierfabrikation* 55 (February 9): 281-282.

1167   ————. 1924. "Württembergische Papiergeschichte." *Wochenblatt für Papierfabrikation* 55 (March 1): 453-457.

1168   ————. 1924. "Württembergische Papiergeschichte." *Wochenblatt für Papierfabrikation* 55 (March 22): 656.

1169   ————. 1924. "Württembergische Papiergeschichte." *Wochenblatt für Papierfabrikation* 55 (August 23): 2132-2136.

1170   ————. 1924. "Württembergische Papiergeschichte." *Wochenblatt für Papierfabrikation* 55 (October 11): 2602-2603.

1171   ————. 1938. *Württembergische Papiergeschichte.* Württemberg, Germany: Grintter-Staib Verlag.

1172   ————. 1930. "Zur Württembergischen Papiergeschichte." *Wochenblatt für Papierfabrikation* 61 (September 27): 1247.

1173   ————. 1927. "Zwei Nachträge Zur Bayerischen Papiergeschichte." *Der Papier-Fabrikant* 25 (November 27): 751-753.

1174   Hülle, F. 1929. "Entwicklung und Bedeutung der Badischen Papiererzeugungsindustrie." Unpublished Ph.D. dissertation, University of Ohlau.

1175   Humulus. 1925. "Geschichte des Papiers." *Wochenblatt für Papierfabrikation* 56 (July 4): 821.

1176   Hunter, Dard. 1943. "An Era in Papermaking—The Story of Dr. Jacob Christian Schäffer." *Journal of the New York Botanical Garden* 44 (July): 149-159.

1177   Jacob, Bruno. 1952. "Die Holzschliff-Verwertung in Kurhessen." *Papiergeschichte* 2 (April): 25-26.

1178   ————. 1950. "Die Papierfabrik Niederkaufungen." *Wochenblatt für Papierfabrikation* 78 (June 15): 308-309.

1179   ————. 1950. "Die Papiermühle Bettenhausen vor Kassel." *Wochenblatt für Papierfabrikation* 78 (February 28): 100-102.

1180   Jacobs, E. 1882. "Alter und Früheste Erzeugnisse der Papierfabrikation in Wernigerode." *Zeitschrift der Harz-Vereins* 15: 141-153.

1181   Jaffé, Albert. 1928. "Die Ehemaligen Papiermühlen im Heutigen Bezirksamt Pirmasens und Ihre Wasserzeichen." *Der Papier-Fabrikant* 26 (September 9): 565-570.

1182   ————. 1928. "Die Neu- Oder Apostelmühle Bei Rodalben." *Der Papier-Fabrikant* 26 (September 16): 581-584.

1183   Kapp, Arno. 1951. "Leipzigs Lumpenhandel und der Papiermangel am Ende des 18. Jahrhunderts in Kursachsen." *Papiergeschichte* 1 (December): 56-57.

1184   Kazmeier, August Wilhelm. 1954. "Druck und Papier des Manifests von Diether von Isenburg von 1462." *Gutenberg-Jahrbuch* 29: 26-35.

1185   ————. 1951. "Historischer Streifzug Durch die Rohstofffragen der Papierherstellung." *Papiergeschichte* 1 (October): 40-45.

1186   ————. 1950. "Der Johannistag 1390 in Nürnberg: ein Beitrag Zur Gründungsfeier der Stadt Nürnberg." *Das Papier* 4 (June): 210-211.

1187   ————. 1957. "Zu Ulman Stromeirs Papiermühlenbericht." *Gutenberg-Jahrbuch* 32: 21-25.

1188   ————. 1954. "Eine Urkunde aus dem Leben Ulman Stromeirs." *Wochenblatt für Papierfabrikation* 82 (August 15): 631.

1189   ————. 1952. "Wasserzeichen und Papier der Zweiundvierzigzeiligen Bibel." *Gutenberg-Jahrbuch* 27: 21-29.

1190   ————. 1951. "Zwei Mailänder Papierdokumente des 14. Jahrhunderts und Ihre Beziehung zu Mainz." *Gutenberg-Jahrbuch* 26: 34-38.

1191   Kehrli, J.O. 1952. "Die Schäffers'schen Papierversuche 1772." *Schweizerishes Gutenbergmuseum* 38: 3-10.

1192   Khaeser, Peter. 1937. "Die Wolfsbronner Papiermühle." *Alt-Gunzenhausen* No. 14: 33-46.

1193   Kieckbusch, Hans. 1957. "Beitrag Zur Geschichte der Papiermühle Schulendorf, eines Alten Industriebetriebes in Gebiet der Früheren Klostergrundherrschaft Ahrensbök." *Papiergeschichte* 7 (December): 81-83.

1194   Killermann-Regensburg, S. 1927. "Jakob Christian Schaeffers Papierversuche (1765-1772)." *Der Papier-Fabrikant* 25 (October 23): 665-670.

1195   ————. 1936. "Jakob Christian Schaeffers Papierversuche 1765-1772." *Sankt Wiborada* 3: 93-96.

1196   Kirschner, E. 1920. "Geschichte. Cröllwitz 1714-1914." *Wochenblatt für Papierfabrikation* 51 (March 27): 854-858.

1197   ————. 1920. "Geschichte. Elsass. Um 1840." *Wochenblatt für Papierfabrikation* 51 (May 22): 1419-1420.

1198   ————. 1920. "Geschichte Notizen." *Wochenblatt für Papierfabrikation* 51 (September 4): 2454.

1199   ————. 1922. "Geschichte. Oberursel, 16.-19. Jahrhundert." *Wochenblatt für Papierfabrikation* 53 (February 11): 495-497.

1200   ————. 1919. "Geschichte. Sachsen-Altenburg. 1816." *Wochenblatt für Papierfabrikation* 50 (February 1): 255.

1201 _____. 1920. "Geschichtliche Notizen." *Wochenblatt für Papierfabrikation* 51 (August 14): 2260.

1202 _____. 1921. "Geschichtliche Notizen." *Wochenblatt für Papierfabrikation* 52 (January 15): 98.

1203 _____. 1920. "Hohenspringe und Schöpsdorf." *Wochenblatt für Papierfabrikation* 51 (March 20): 787.

1204 _____. 1920. "Papiermühle Neustadt i. M." *Wochenblatt für Papierfabrikation* 51 (March 20): 787.

1205 _____. 1898. "Die Papiermühlen Bei Chemnitz." *Papier-Zeitung* 23 (November 13): 3443.

1206 _____. 1920. "Sachsen und Schlesien." *Wochenblatt für Papierfabrikation* 51 (December 24): 3620.

1207 Kleeberger, Karl. 1925. "Urkundliches Über die Papiermühle in Mosbach." *Mannheimer Geschichtsblatter* 26: 10-15.

1208 Klingelschmitt, Franz Theodor. 1936. "Eine Kurmainzer Papiermühle." *Philobiblon* 9 (September-October): 300-302.

1209 _____. 1940. "Kurmainzer Papiermühlen." *Philobiblon* 12 (March): 96-104.

1210 Knopff, Max. 1951. "Das Cospudener Papiermühlenprivileg." *Wochenblatt für Papierfabrikation* 79 (January 15): 10.

1211 Koch, Herbert. 1956. "Aus der Schreibstube des Städtrates in Leipzig 1475-1500." *Gutenberg-Jahrbuch* 31: 54-56.

1212 Kohtz, Hans. 1954. "Die Herkunft Ostpreussischer Papiermacher." *Papiergeschichte* 4 (February): 2-9.

1213 _____. 1951. "Ostpreussische Adler-Wasserzeichen im Wandel der Zeit." *Papiergeschichte* 1 (December): 47-53.

1214 Körholz, Franz. 1952. "Die Papiermühle zu Scheppen." *Papiergeschichte* 2 (November): 65-72.

1215 Kühne, H. 1964. "Das Schicksal der Ersten Wittenberger Papiermühle in der Mitte des 16. Jahrhunderts." *Wochenblatt für Papierfabrikation* 92 (June): 350-351.

1216 Kühnert, Herbert. 1956. "Über eine im Jahre 1451 Bei der Stadt Coburg Gegründete Papiermühle." *Papiergeschichte* 6 (July): 34-38.

1217 Kunze, Horst. 1950. "Das 'Papier' in den Deutschen Allgemeinen-zyklopadien des 18. und 19. Jahrhunderts." *Gutenberg-Jahrbuch* 25: 41-45.

1218 Labarre, E.J., and H. Voorn. 1949. "Les Loisirs des Boids du Loing." *De Papierwereld* 4 (September): 70-72.

1219 Layer, Adolf. 1967. "Allgäuer Papiermacher in der Fremde." *Gutenberg-Jahrbuch* 42: 14-16.

1220 Loeber, E.G. 1970. "Zu L.C. Sturm's Papiermühlenrissen."
   Papiergeschichte 20 (December): 49-72.
1221 _____. 1972. "Die Plögersche Papiermühle in Schieder."
   Papiergeschichte 22 (June): 19-26.
1222 Lucke, C. 1925. "Geschichte des Papiers." *Wochenblatt für*
   Papierfabrikation 56 (May 23): 644.
1223 Marabini, Edmund V. 1937. "Bayerische Papiergeschichte." *Der*
   Papier-Fabrikant 35 (December 3): 493-496.
1224 Martell, P. 1927. "Zur Geschichte der Papierindustrie in Schlesien."
   Der Papier-Fabrikant 25 (June A): 36-42.
1225 _____. 1926. "Zur Geschichte der Papierindustrie in Westfalen."
   Der Papier-Fabrikant 24 (October 3): 609-613.
1226 Meldau, Robert. 1937. "Reichsprivilegien für Wasserzeichen."
   Gutenberg-Jahrbuch 12: 13-17.
1227 Merkel, Johannes. 1927. "Aus der Schlesischen Papiergeschichte.
   Die 3 Papiermühlen am Oberlauf des Queis." *Wochenblatt für*
   Papierfabrikation 58 (June 18A): 10-14.
1228 Metzger, H. 1910. "Geschichte der Papiermühle zu Friedland."
   Mitteilungen der Vereins für Geschichte der Deutschen in
   Böhmen 48: 302-345.
1229 Meyer, F.H. 1888. "Papierfabrikation und Papierhandel. Beiträge
   zu Ihrer Geschichte, Besonders in Sachsen." *Archiv für*
   Geschichte des Deutschen Buchhandels 11: 283-357.
1230 Michaelsen, Hermann. 1955. "Die Kasseedorfer Papiermühle."
   Papiergeschichte 5 (November): 66-68.
1231 Mitterwieser, Alois. 1937. "Alte Papiermühlen des Adels in
   Südbayern." *Gutenberg-Jahrbuch* 12: 9-12.
1232 _____. 1938. "Die Alten Papiermühlen Bei Landsberg am Lech."
   Sankt Wiborada 5: 90-93.
1233 _____. 1940. "Die Alten Papiermühlen Münchens." *Gutenberg-*
   Jahrbuch 15: 25-34.
1234 _____. 1939. "Die Alten Papiermühlen von Landshut an der Isar
   und Braunau am Inn." *Gutenberg-Jahrbuch* 14: 31-38.
1235 _____. 1933. "Frühere Papiermühlen in Altbayern und Ihre
   Wasserzeichen." *Gutenberg-Jahrbuch* 8: 9-22.
1236 _____. 1938. "Geschichte der Papiermühle zu Lengfelden Bei
   Salzburg." *Der Papier-Fabrikant* 36 (November 25): 485-487.
1237 _____. 1942-1943. "Papierpreise in Früheren Zeiten." *Buch und*
   Schrift 5-6: 159-174.
1238 Moellendorff, Moritz Ulrich. 1932. "Alter Traum und Frühe

Erfüllung. Ein Kuriosum aus der Geschichte der Papiermacherei." *Zeitschrift für Buchfreunde (Third Series)* 1 (July): 153-155.

1239   Moessner, Gustav. 1967. "Papiermühle in Bad Reinerz." *Papiergeschichte* 17 (November): 70.

1240   Nadler, Alfred. 1963. "Die Papierhandels-Compagnie Wasmann zu Würzburg und Ihre Wasserzeichen (1773-1818)." *Papiergeschichte* 13 (December): 74-76.

1241   _____. 1962. "Die Verwendung Holländischer Papiere in Mainfranken Nach 1700 und Nachbildungen Holländischer Wasserzeichen in Suddeutschland." *Papiergeschichte* 12 (February): 20-21.

1242   Neder, Emil. 1905. "Die Papiermühle zu Bensen 1569-1884." *Mitteilungen der Vereins für Geschichte der Deutschen in Böhmen* 44: 220-234.

1243   Oechelhäuser, A. 1941. "Die Erfindung der Selbstabnahmemaschine." *Wochenblatt für Papierfabrikation* 72 (December 6): 706-709.

1244   Pabich, Franciszek. 1965. "Dzieje Najstarszej Papierni w Prusach Królewskich." *Przegląd Papierniczy* 21 (December): 387-389.

1245   Paffrath, Arno. 1966. "Die Papiermühlen in Leppetal. Beiträge Zur Geschichte der Papiermühlen im Bergischen Lande." *Papiergeschichte* 16 (April): 19-25.

1246   Paper Publications Society. 1956. *The Nostitz Papers.* Hilversum, The Netherlands: The Paper Publications Society.

1247   Pels, C. 1949. "Iets Naders Over Ulman Stromer." *De Papierwereld* 4 (August): 48.

1248   Peschke, Werner. 1937. "Die Mühlen der Mark im Mittelalter und Ihre Technische Einrichtung." *Brandenburgishe Jahrbücher* 6: 55-59.

1249   Piccard, Gerhard. 1953. "Die Älteste Deutsche Papiermühle: die Papierzeichen der Gleissmühle Bei Nürnberg." *Wochenblatt für Papierfabrikation* 81 (July 15): 486-490.

1250   _____. 1961. "Zur Geschichte der Papiererzeugung in der Reichsstadt Memmingen." *Archiv für Geschichte des Buchwesens* 3: 595-612.

1251   _____. 1963. "Rechtsrheinische (Badische) Papiermühlen und Ihre Beziehungen zu Strassburg." *Archiv für Geschichte des Buchwesens* 4: 997-1102.

1252   _____. 1962. "Über die Anfänge des Gebrauchs des Papiers in Deutschen Kanzleien." *Studi in Onore di Amintore Fanfani* n.v.: 347-401.

1253 _____. 1967. "Das Wappen der Herzöge von Jülich-Berg, Cleve, Mark und Ravensberg." *Papiergeschichte* 17 (May): 1-14.

1254 _____, and Lore Sporhan-Krempel. 1950. "Beiträge Zur Badischen Papiergeschichte: Urkundliches Material Über die Baukosten von Papiermühlen, Papierproduktion und Papiererlöhne im 18. Jahrhundert in den Badischen Stammlanden und Vorderösterreich." *Wochenblatt für Papierfabrikation* 78 (June 30): 357-360.

1255 _____, and _____. 1950. "Kloster Salem und Papiere von Ravensburg vor und Nach dem 30jahrigen Krieg." *Das Papier* 4 (June): 208-210.

1256 _____, and _____. 1951. "Die Untersuchung der Ältesten Erhaltenen Zeitungen auf Ihre Wasserzeichen." *Papiergeschichte* 1 (May): 13-14.

1257 Piette-Rivage, Ludwig von. 1969. "Geschichte der Papiermacherfamilie Piette de Rivage." *Papiergeschichte* 19 (November): 25-28.

1258 Raithelhuber, Ernst. 1966. "Heinrich Voelter und die Entwicklung des Holzschleifers." *Papiergeschichte* 16 (December): 1-18.

1259 _____. 1971. "Die Konstruktionen der Ersten Papiermaschinen Mit Geschütteltem Langsieb." *Papiergeschichte* 21 (October): 22-32.

1260 _____. 1971. "Die Konstruktionen der Ersten Papiermaschinen Mit Geschütteltem Langsieb." *Papiergeschichte* 21 (December): 33-52.

1261 Rauchheld, Hans. 1928. "Oldenburgische Papiermühlengeschichte." *Wochenblatt für Papierfabrikation* 59 (February 18): 172-173.

1262 _____. 1926. "Die Wasserzeichen der Ersten in Oldenburg Gedruckten Bücher." *Der Papier-Fabrikant* 24 (October 24): 663-665.

1263 Renker, Armin. 1935. "Alte Papiermühlen Einst und Jetzt." *Buch und Werbekunst* 12: 96-98.

1264 _____. 1950-1951. "Vom Brauchtum der Alten Papiermacher." *Imprimatur* 10: 115-120.

1265 _____. 1937. "Eine Briefmarke Mit dem Bild des Papiermachers." *Der Papier-Fabrikant* 35 (December 10): 510-511.

1266 _____. 1960. "Vom Büttenpapier Einst und Heute." *Allgemeine Papier-Rundschau* No. 6 (March 20): 280-283.

1267 _____. 1954. "La Feuille Blanche. Handpapiermacherei in Frankreich." *Gutenberg-Jahrbuch* 29: 21-25.

1268 _____. 1958. "Jacob Christian Schäffer in der Geschichte der Papierherstellung." *Gutenberg-Jahrbuch* 33: 30-36.

1269 _____. 1950. "Von den Leichen im Wasser: eine Papiermacher-Mär." *Allgemeine Papier-Rundschau* No. 3 (February 15): 89.

1270 _____. 1961. "Moritz Friedrich Illig 1777-1845: Inventor of Rosin Sizing." *The Paper Maker* 30 (September): 37-43.

1271 _____. 1938. "Mussestunden an den Ufern des Lorng." *Der Papier-Fabrikant* 36 (May 27): 186-188.

1272 _____. 1936. "Ein Nützliches Mühlwerk." *Wochenblatt für Papierfabrikation* 67 (September 26): 726-727.

1273 _____. 1959. "Pioneer in the Development of Wood Pulp: Jacob Christian Schäffer." *The Paper Maker* 28 (September): 19-23.

1274 _____. 1938. "Von Schäffer zu Keller." *Der Papier-Fabrikant* 36 (July 8): 317-324.

1275 _____. 1938. "Von Schäffer zu Keller." *Der Papier-Fabrikant* 36 (July 1): 311-316.

1276 _____. 1961. "Some Curious Customs of Old-Time Papermaking in Germany." *The Paper Maker* 30 (February): 3-11.

1277 _____. 1938. "Eine Warnung Vor dem 'Hölzernen Papier'." *Der Papier-Fabrikant* 36 (September 9): 395.

1278 Rirchhoff, Ulbrecht. 1879. "Deutschen Papierhandel im Beginn des 18. Jahrhunderts." *Archiv für Geschichte des Deutschen Buchhandels* 2: 254-257.

1279 Ruppel, A. 1937. "Die Erfindung Gutenbergs, Technisch und Geistig Gesehen." *Wochenblatt für Papierfabrikation* 68 (Special Edition): 55-56.

1280 _____. 1938. "Wann Starb Gutenberg? Wo Liegt Er Begraben?" *Wochenblatt für Papierfabrikation* 69 (Special Edition): 15-16.

1281 Schacht, Gust. 1911. "Die Papiere und Wasserzeichen des XIV. Jahrhunderts im Hauptstaatsarchiv zu Dresden." *Wochenblatt für Papierfabrikation* 42 (June 10): 2127-2137.

1282 Schlieder, Wolfgang. 1961. "Bemerkungen Zur Einführung der Papiermaschine in Deutschland." *Papiergeschichte* 11 (December): 82-84.

1283 _____. 1970. "Die Einführung der Papiermaschine in Deutschland." *Jahrbuch der Deutschen Bücherei* 6: 101-126.

1284 _____. 1966. "Zur Geschichte der Papierherstellung in Deutschland. Von dem Anfängen der Papiermacherei Bis Zum 17. Jahrhundert." *Beiträge Zur Geschichte des Buchwesens* 2: 33-168.

1285   ————. 1972. "Historia Papiernictiva i Badania Historyczne w Zakładach Papierniczych w Niemieckiej Republice Demokratycznej." *Przegląd Papierniczy* 28 (March): 98-100.

1286   Schmidt, A. 1925. "Geschichte der Wittenberger Papiermühlen." *Zeitschrift für Buchkunde* 2: 19-32.

1287   Schmidt, R. 1928. "Märkische Papiermühlen Bis um 1800." *Brandenburgische Jahrbücher* 3: 58-76.

1288   Schmitz, F. 1931. "Die Älteste Papiermühle zu Bergisch Gladbach." *Wochenblatt für Papierfabrikation* 62 (July 11): 665-667.

1289   Schreiber, J. 1929. "Geschichte der Papiermühlen zu Rokitnitz, Zugleich ein Beitrag Zur Stadtgeschichte." *Mitteilungen der Vereins für Geschichte der Deutschen in Böhmen* 67: 87-114.

1290   Schulte, Alfred. 1931. "Das Alte Lied der Papiermacher." *Wochenblatt für Papierfabrikation* 62 (July 18): 688.

1291   ————. 1932. "Die Ältesten Papiermühlen der Rheinlande." *Gutenberg-Jahrbuch* 7: 44-52.

1292   ————. 1941. "Die Anfänge der Buntpapierherstellung in Deutschland." *Archiv für Buchgewerbe und Gebrauchsgraphik* 78 (June): 223-225.

1293   ————. 1939. "Die Anfänge des Deutschen Papiermaschinenbaues." *Der Papier-Fabrikant* 37 (April 28): 149-151.

1294   ————. 1933. "Das Ehrliche Geschenk und der Willkomm der Papierer." *Wochenblatt für Papierfabrikation* 64 (Special Edition): 5-13.

1295   ————. 1936. "Vom Ehrlichen Geschenk." *Wochenblatt für Papierfabrikation* 67 (Special Edition): 52-54.

1296   ————. 1934. "Entrümpelung und Papiergeschichte." *Wochenblatt für Papierfabrikation* 65 (May 12): 335-336.

1297   ————. 1938. "Die Entwicklung der Rheinischen Papiermacherei." *Der Papier-Fabrikant* 36 (December): 538-540.

1298   ————. 1931. "Zur Entwicklungsgeschichte der Schattierten Wasserzeichen." *Wochenblatt für Papierfabrikation* 62 (June 6A): 18-23.

1299   ————. 1933. "Der Erste Holländer im Harz." *Der Papier-Fabrikant* 31 (July A): 77-79.

1300   ————. 1932. "Die Geschichtlichen Zahlen der Deutschen Bütten und Papiermaschinen." *Wochenblatt für Papierfabrikation* 63 (April 9): 282-283.

1301   ————. 1939. "Die Gesellenprüfung im Alten Handwerk." *Wochenblatt für Papierfabrikation* 70 (June 3): 465-467.

1302 _____. 1934. "Die Gestohlene Uhr Oder: Viel Lärm um Nichts. Eine Papiermacher-Scheltensache der 1770er Jahre aus der Lausitz." *Wochenblatt für Papierfabrikation* 65 (Special Edition): 41-46.

1303 _____. 1930. "Johann Widmann in Heilbronn, der Erste Deutsche Papiermaschinenfabrikant." *Wochenblatt für Papierfabrikation* 61 (June 21A): 17-23.

1304 _____. 1937. "Von Kleidung und Tracht der Alten Papiermacher." *Wochenblatt für Papierfabrikation* 68 (November 27): 931-934.

1305 _____. 1938. "Von Kleidung und Tracht der Alten Papiermacher." *Wochenblatt für Papierfabrikation* 69 (January 1): 10-12.

1306 _____. 1935. "Krause Geschichten aus Kleinen Betrieben." *Wochenblatt für Papierfabrikation* 66 (Special Edition): 51-55.

1307 _____. 1937. "Von der Lumkorherei in Früherer Zeit." *Der Papier-Fabrikant* 35 (December 22): 518-520.

1308 _____. 1941. "Die Papiermacherei in Biberach an der Riss." *Wochenblatt für Papierfabrikation* 72 (July 5): 385-387.

1309 _____. 1937. "Die Papiermühle zu Boisenburg a.d. Elbe." *Wochenblatt für Papierfabrikation* 68 (November 20): 890.

1310 _____. 1937. "Die Papiermühle zu Jettenbach am Inn." *Wochenblatt für Papierfabrikation* 68 (February 6): 102-103.

1311 _____. 1937. "Die Papiermühle zu Jettenbach am Inn." *Wochenblatt für Papierfabrikation* 68 (December 18): 1005.

1312 _____. 1936. "Die Papiermühle zu Kiedrich im Rheingau." *Wochenblatt für Papierfabrikation* 67 (July 11): 521-522.

1313 _____. 1953. "Ravensburg." *Papiergeschichte* 3 (April): 13-26.

1314 _____. 1936. "Die Riesumschlagdrucke der Papiermacher." *Gutenberg-Jahrbuch* 11: 14-22.

1315 _____. 1936. "Tagebuch Eines Wandernden Gesellen." *Wochenblatt für Papierfabrikation* 67 (May 2): 337-338.

1316 _____. 1936. "Tagebuch Eines Wandernden Gesellen." *Wochenblatt für Papierfabrikation* 67 (May 23): 394-395.

1317 _____. 1941. "Über das Feuchten des Papiers Mit Nassen Tüchern Bei Johann Zainer und Einigen Anderen Frühdruckern." *Gutenberg-Jahrbuch* 16: 19-22.

1318 _____. 1931. "Ueber die Erste Papiermaschine und Ihre Erfindung." *Der Papier-Fabrikant* 29 (November 22): 747-749.

1319 _____. 1933. "Woher Stammt Unser Wappen?" *Wochenblatt für Papierfabrikation* 64 (December 16): 905-906.

1320 _____. 1936. "Zwei Papiermühlen Bei St. Goarshausen."
Wochenblatt für Papierfabrikation 67 (December 5): 919-920.

1321 Schulte, Toni. 1956. "Alte Augsburger Modellbaukunst des 18.
Jahrhunderts." *Papiergeschichte* 6 (November): 66-68.

1322 _____. 1957. "Zur Ältesten Darstellung Eines Rührwerkes."
Papiergeschichte 7 (December): 75-80.

1323 _____. 1957. "Krankheiten und Unfälle in der Alten
Handpapierindustrie." *Papiergeschichte* 7 (October): 69-74.

1324 _____. 1952. "Eine Papiermacher-Ordung aus dem Jahre 1546."
Papiergeschichte 2 (June): 36-41.

1325 _____. 1958. "Die Spindelpresse Bei Salmon De Caus."
Papiergeschichte 8 (December): 79-81.

1326 _____. 1955. "Die Zeitung und Ihr Papier Von Cäsar Über
Gutenberg Bis Zur Neuzeit." *Papiergeschichte* 5 (November):
68-69.

1327 Schwanke, Erich. 1920. "Beitrag Zur Geschichte der Papiermacherei
in Deutschböhmen." *Wochenblatt für Papierfabrikation* 51
(November 13): 3175-3177.

1328 Schwann, Mathieu. 1924. "Aus Alten Zeit Deutschen
Papiermacherei." *Der Papier-Fabrikant* 22 (November 2):
519-522.

1329 _____. 1926. "Zur Entwicklung der Rheinischen Papierindustrie."
Der Papier-Fabrikant 24 (January 10): 24-26.

1330 Schwartz, Hubertus. 1954. "Die Papiermühle Ardey Vor Soest."
Papiergeschichte 4 (December): 69-70.

1331 Spahr, W. 1927. "Altenburg und die Papiermacher." *Wochenblatt
für Papierfabrikation* 58 (January 29): 111-114.

1332 Sporhan-Krempel, Lore. 1953. "Die Älteste Deutsche Papiermühle:
die Geschichte der Gleissmühle zu Nürnberg." *Wochenblatt
für Papierfabrikation* 81 (July 15): 484-486.

1333 _____. 1950. "Die Badischen Papiermühlen und Ihre
Wasserzeichen." *Das Papier* 4 (September): 351-353.

1334 _____. 1958. "Das Ende der Ravensburger Papiermacherei."
Wochenblatt für Papierfabrikation 86 (February 28): 126-128.

1335 _____. 1956. "Franz Xaver Bullinger, Papierfabrikant zu
Unterkochen." *Wochenblatt für Papierfabrikation* 84 (April
15): 231-232.

1336 _____. 1959. "Frühe Verwendung Von 'Gewerblichem' Papier."
Papiergeschichte 9 (December): 74-76.

1337 _____. 1953. "Die Frühzeit der Papiermacherei in Ravensburg."
Wochenblatt für Papierfabrikation 81 (March 31): 196-199.

1338   _____. 1950. "Aus der Geschichte der Badischen Papiermühlen und Ihrer Wasserzeichen." *Wochenblatt für Papierfabrikation* 78 (December 15): 696-700.

1339   _____. 1964. "Geschichte der Papierfabrikation im Filstal." *Archiv für Geschichte des Buchwesens* 5: 461-522.

1340   _____. 1960. "Die Geschichte der Papierindustrie im Kreis Wangen." *Archiv für Geschichte des Buchwesens* 2: 130-148.

1341   _____. 1955. "Zur Geschichte der Papierwerke im Kreis Wangen." *Wochenblatt für Papierfabrikation* 83 (June 30): 507-511.

1342   _____. 1955. "Zur Geschichte der Papierwerke im Kreis Wangen: 2. Die Papiermühle zu Niederwangen." *Wochenblatt für Papierfabrikation* 83 (October 31): 826-828.

1343   _____. 1972. "Handel und Händler Mit Ravensburger Papier." *Papiergeschichte* 22 (June): 29-36.

1344   _____. 1972. "Handel und Händler Mit Ravensburger Papier." *Papiergeschichte* 22 (December): 37-40.

1345   _____. 1960. "Die 'Legende' von der Heilmannschen Papiermühle zu Strasburg." *Papiergeschichte* 10 (December): 71-74.

1346   _____. 1949. "Der Luftschiffer Blanchard und die Schwäbischen Papiermacher." *Wochenblatt für Papierfabrikation* 77 (September): 341.

1347   _____. 1949. "150 Jahre Papierindustrie im Strudelbachtel." *Wochenblatt für Papierfabrikation* 77 (August): 267-268.

1348   _____. 1956. "Die Papierindustrie in Baden-Württemberg Zur Zeit der Einführung der Papiermaschine." *Wochenblatt für Papierfabrikation* 84 (July 31): 573-575.

1349   _____. 1967. "Papiermacherei im Ehemaligen Hochstift Osnabrück." *Archiv für Geschichte des Buchwesens* 8: 333-390.

1350   _____. 1957. "Die Papiermühle des Klosters Weingarten zu Albisreute 'im Loch'." *Wochenblatt für Papierfabrikation* 85 (April 15): 250-252.

1351   _____. 1957. "Die Papiermühle zu Berg Bei Stuttgart." *Wochenblatt für Papierfabrikation* 85 (March 31): 214.

1352   _____. 1954. "Die Papiermühle zu Ulm Bis Zum Ende der Reichsstadtzeit." *Wochenblatt für Papierfabrikation* 82 (June 15): 467-469.

1353   _____. 1966. "Die Papiermühle zu Unterkochen." *Archiv für Geschichte des Buchwesens* 6: 547-574.

1354 _____. 1966. "Die Papiermühle Zum Doos." *Archiv für Geschichte des Buchwesens* 6: 1309-1324.

1355 _____. 1957. "Die Papiermühlen Bei Lindau von 1650-1850." *Wochenblatt für Papierfabrikation* 85 (September 30): 707-710.

1356 _____. 1957. *Papiermühlen und Papiermacher in Lindau und Oberschwaben.* Lindau, Germany: J. Thorbecke.

1357 _____. 1966. "Die Papiermühlen zu Ankum im Hochstift Osnabrück." *Papiergeschichte* 16 (October): 4-7.

1358 _____. 1953. "Die Papiermühlen zu Enzberg." *Papiergeschichte* 3 (February): 4-8.

1359 _____. 1957. "Die Papiermühlen zu Lindau." *Wochenblatt für Papierfabrikation* 85 (June 30): 476-478.

1360 _____. 1964. "Die Papierrechnungen von Johann Friedrich Cotta 1788-1806. Mit Einem Kommentar." *Archiv für Geschichte des Buchwesens* 5: 1369-1472.

1361 _____. 1952. "Das Papierwerk in Frankeneck/Pfalz." *Papiergeschichte* 2 (September): 55-58.

1362 _____. 1960. "Die Papierwirtschaft der Nürnberger Kanzlei und die Geschichte der Papiermacherei im Gebiet der Reichsstadt Bis Zum Beginn des 30jährigen Krieges." *Archiv für Geschichte des Buchwesens* 2: 161-169.

1363 _____. 1954. "Die Soziale Lage der Papierergesellen zu Ravensburg Gegen Ende des 18. Jahrhunderts." *Wochenblatt für Papierfabrikation* 82 (December 15): 976-977.

1364 _____. 1958. "Der Versuch Zur Erlangung Eines Inländischen Papiermonopols im Herzogtum Württemberg." *Wochenblatt für Papierfabrikation* 86 (September 30): 808-810.

1365 _____. 1973. "Vier Jahrhunderte Papiermacherei in Reutlingen (ca. 1465-1863)." *Archiv für Geschichte des Buchwesens* 13: 1513-1586.

1366 _____. 1954. "Das Wasserzeichen in der Sicht des Wirtschaftsgeschichtsforschers und des Archivars." *Wochenblatt für Papierfabrikation* 82 (December 15): 978.

1367 _____. 1957. "Die Wasserzeichenkartei im Württembergischen Hauptstaatsarchiv zu Stuttgart." *Philobiblon (Hamburg)* 1 (September): 231-232.

1368 _____. 1949. "Zur Württembergischen Papiergeschichte." *Wochenblatt für Papierfabrikation* 77 (November): 452.

1369 _____, and Gerhard Piccard. 1952. "Die Papiermühle zu Berg Bei Stuttgart." *Papiergeschichte* 2 (April): 26-28.

1370   _____, and _____. 1951. "Die Papiermühle zu Niefern."
    Papiergeschichte 1 (August): 17-24.

1371   _____, and _____. 1951. "Der Versuch Zur Erlangung Eines
    Inländischen Papiermonopols im Herzogtum Württemberg."
    Papiergeschichte 1 (December): 53-56.

1372   _____, and Wolfgang von Stromer. 1963. "Die Früheste
    Geschichte Eines Gewerblichen Unternehmens in Deutschland:
    Ulman Stromers Papiermühle in Nürnberg. Mit Einem
    Wasserzeichengutachten von Gerhard Piccard." *Archiv für
    Geschichte des Buchwesens* 4: 187-212.

1373   _____, and _____. 1960. "Das Handelshaus der Stromer von
    Nürnberg und die Geschichte der Ersten Deutschen
    Papiermühle." *Vierteljahresschrift für Sozial- und
    Wirtschaftsgeschichte* 47 (March): 81-104.

1374   Springer, Karl. 1928. "Ettlinger Wasserzeichen." *Badische Heimat*
    15: 232-239.

1375   Steinbrucker, Charlotte. 1957. "Vom Schreib- und Briefpapier."
    Wochenblatt für Papierfabrikation 85 (March 15): 174.

1376   Steiner, Gerhard. 1938. "Geschichte der Papiermühle Sachsendorf i.
    Thür." *Wochenblatt für Papierfabrikation* 69 (September 24):
    799-803.

1377   _____. 1939. "Der Lehrbraten der Papiermacher." *Wochenblatt
    für Papierfabrikation* 70 (June 3): 467-469.

1378   Steinmüller Karl. 1964. "Der Nachloss Eines Papiermachergesellen."
    Papiergeschichte 14 (April): 1-5.

1379   _____. 1957. "Wanderbücher von Papiermachern im Stadtarchiv
    Zwikau." *Papiergeschichte* 7 (October): 59-69.

1380   Stieda, Wilhelm. 1915. "Mecklenburgische Papiermühlen."
    *Jahrbücher des Vereins für Mecklenburgische Geschichte und
    Altertumskunde* 80: 115-184.

1381   Stromer-Reichenbach, Wolfgang von, and Lore Sporhan-Krempel.
    1963. "Die Erste Deutsche Papiermühle." *Papiergeschichte* 13
    (December): 67-74.

1382   Tacke, Eberhard. 1954. "Beiträge Zur Geschichte des Papiers in
    Niedersachsen und Angrenzenden Gebieten." *Papiergeschichte*
    4 (July): 35-44.

1383   _____. 1954. "Beiträge Zur Geschichte des Papiers in
    Niedersachsen und Angrenzenden Gebieten (II)."
    Papiergeschichte 4 (December): 71-79.

1384   _____. 1955. "Beiträge Zur Geschichte des Papiers in

Niedersachsen und Angrenzenden Gebieten (II)."
Papiergeschichte 5 (December): 79-82.

1385 ————. 1956. "Beiträge Zur Geschichte des Papiers in
Niedersachsen und Angrenzenden Gebieten (IV)."
Papiergeschichte 6 (December): 73-75.

1386 ————. 1968. "Eine Bisher Unbekannte Papiermühle des Frühen
17. Jahrhunderts Bei Salzdahlum (Kreis Wolfenbüttel)."
Papiergeschichte 18 (July): 52.

1387 ————. 1960. "Zur Entstehungs- und Frühgeschichte der
Papiermühle Oker." *Braunschweigische Heimat* 46: 44-50.

1388 ————. 1965. "Von den Papiermachern Zur Grane und Ihren
Wasserzeichen." *Harz-Zeitschrift* 17: 79-104.

1389 ————. 1957. "Die Papiermühle Neddenaverbergen-Kükenmoor."
Papiergeschichte 7 (April): 13-18.

1390 ————. 1957. "Die Papiermühle Neddenaverbergen-Kükenmoor."
Papiergeschichte 7 (May): 29-37.

1391 ————. 1965-1966. *Die Schaumburger Papiermühlen und Ihre
Wasserzeichen im Rahmen der Nordwestdeutschen
Papiergeschichte.* Bückeburg, Germany: Verlag Grimme
Bückeburg. 2 Volumes.

1392 ————. 1964. "Standorte der Papiererzeugung in Niedersachsen
und Angrenzenden Gebieten." *Neues Archiv für Niedersachsen
und Angrenzende Gebiete* 13 (December): 253.

1393 ————. 1959. "Eine Weitere Papiermühle des Ausgehenden 16.
Jahrhunderts am Westharz Bei Gittelde." *Papiergeschichte* 9
(October): 48-50.

1394 ————, and Irmgard Tacke. 1967. "Über Sortenwasserzeichen
des 18. Jahrhunderts in Niedersachsen." *Papiergeschichte* 17
(November): 62-65.

1395 Thiel, Viktor. 1934. "Die Geschichte Bedeutung der Schwäbischen
Papiererzeugung." *Wochenblatt für Papierfabrikation* 65
(Special Edition): 47-51.

1396 ————. 1937. "Zur Geschichte der Papiererzeugung in Tirol,
Vorarlberg und in den 'Vorlanden'." *Wochenblatt für
Papierfabrikation* 68 (Special Edition): 9-19.

1397 ————. 1938. "Zur Geschichte der Papiererzeugung in Tirol,
Vorarlberg und in den 'Vorlanden'." *Wochenblatt für
Papierfabrikation* 69 (April 30): 384-385.

1398 ————. 1938. "Zur Geschichte der Papiererzeugung in Tirol,
Vorarlberg und in den 'Vorlanden'." *Wochenblatt für
Papierfabrikation* 69 (May 28): 468-469.

1399 _____. 1938. "Zur Geschichte der Papiererzeugung in Tirol, Vorarlberg und in den 'Vorlanden'." *Wochenblatt für Papierfabrikation* 69 (June 11): 507-509.

1400 _____. 1939. "Die Geschichte Einer Deutschen Papiererfamilie." *Wochenblatt für Papierfabrikation* 70 (February 11): 127-129.

1401 _____. 1939. "Die Geschichte Einer Deutschen Papiererfamilie." *Wochenblatt für Papierfabrikation* 70 (February 18): 151-153.

1402 _____. 1940. "Das Geschichtliche Wirken der Deutschen Papiererzeugung Nach dem Südosten." *Gutenberg-Jahrbuch* 15: 35-40.

1403 _____. 1932. "Papiererzeugung und Papierhandel Vornehmlich in den Deutschen Landen von den Ältesten Zeiten Bis Zum Beginn des 19. Jahrhunderts." *Archivalische Zeitschrift* 41: 106-151.

1404 _____. 1935. "Schwäbische Einflüsse auf die Entwicklung der Papiererzeugung in den Österreichischen Ländern." *Vierteljahrsschrift für Sozialund Wirtschaftsgeschichte* 38: 282-286.

1405 _____. 1934. "Ueber das Aufkommen von Papierverwendung auf Deutschen Boden." *Der Papier-Fabrikant* 32 (April 15): 175-176.

1406 Vock, Walther E. 1935. "Ein Vertrag der Allgäuer Papiermüller vom 19. Juli 1586." *Allgäuer Geschichtsfreund (New Series)* 38: 45-48.

1407 Volf, J. 1929. "Aus der Geschichte der Papiermühle in Weisswasser Unterm Bösig." *Der Papier-Fabrikant* 27 (January 13): 23-27.

1408 Voorn, Henk. 1954. "Alexander Mitscherlich: Inventor of Sulphite Wood Pulp." *The Paper Maker* 23 (February): 41-44.

1409 _____. 1949. "Naschrift van de Redactie." *De Papierwereld* 4 (August): 48-49.

1410 _____. 1953. "Random Rambles Through the Glorious Past of German Papermaking." *The Paper Maker* 22 (February): 1-9.

1411 Wagner, Karl. 1921. "Die Bevölkerungsbewegung des Kirchspiels Gegenbach im 17. und 18. Jahrhundert." Unpublished Ph.D. dissertation, University of Freiburg.

1412 Weckbach, Hubert. 1965. "Zur Geschichte der Heilbronner Papiermühlen." *Papiergeschichte* 15 (June): 45-52.

1413 _____. 1965. "Zur Geschichte der Heilbronner Papiermühlen." *Papiergeschichte* 15 (December): 53-59.

1414 Weiss, Karl Theodor. 1940. "Abschied vom Papierwerk Gegenbach." *Wochenblatt für Papierfabrikation* 71 (October 26): 561-568.

1415   _____ . 1942. "Ein Besuch in der Frankfurter Papiermühle
        1587." *Wochenblatt für Papierfabrikation* 73 (June 20): 195.
1416   _____ . 1932. "Das Bild des Papierers. Eine Illustration zu
        Goethes Dichtung und Wahrheit." *Zeitschrift für Büchfreunde
        (Third Series)* 1 (July): 145-146.
1417   _____ . 1916. "The German Armorial Watermarks." *Paper* 18
        (July 12): 12-13.
1418   _____ . 1948. "Das Greizer Schlängle." *Das Papier* 2 (October):
        373-375.
1419   _____ . 1921. "Ein Papiermacherscherz." *Wochenblatt für
        Papierfabrikation* 52 (August 24): 2670-2671.
1420   _____ . 1942. "Papier- und Pergamentstreit." *Wochenblatt für
        Papierfabrikation* 73 (December 26): 427-428.
1421   _____ . 1951. "Das Papierwerk zu Gengenbach Seine Geschichte
        und Seine Wasserzeichen." *Die Ortenau* 31: 1-51.
1422   _____ . 1952. "Das Papierwerk zu Gengenbach Seine Geschichte
        und Seine Wasserzeichen." *Die Ortenau* 32: 111-176.
1423   _____ . 1955. "Eine Versteckte Papiermühlengeschichte."
        Wochenblatt für Papierfabrikation 83 (December 15): 983.
1424   _____ . 1929. "Württembergische Papiergeschichte."
        Familiengeschichtliche Blätter 27: 131-132.
1425   Weiss, Wisso. 1949. "Von der Alten Papiermacherei: Unter Welchen
        Voraussezungen Konnte ein Papiermachergeselle Meister
        Werden?" *Wochenblatt für Papierfabrikation* 77 (November):
        424-426.
1426   _____ . 1957. "Eine Angebliche Windpapiermühle."
        Papiergeschichte 7 (December): 83-86.
1427   _____ . 1974. "Zu Einigen Papiermacher-Pokalen."
        Papiergeschichte 24 (November): 6-16.
1428   _____ . 1953. "Das Fortuna - Wasserzeichen." *Wochenblatt für
        Papierfabrikation* 81 (December 15): 901-906.
1429   _____ . 1949. "Goethe Als Papierliebhaber." *Wochenblatt für
        Papierfabrikation* 77 (October): 367-368.
1430   _____ . 1966. "Zur Lumpensammelkonzession für den
        Buchdrucker Hieronymus Joh. Struck in Stralsund."
        Gutenberg-Jahrbuch 41: 18-25.
1431   _____ . 1967. "Das Lumpensammelprivileg für die Buchdrucker
        Struck in Stralsund." *Gutenberg-Jahrbuch* 42: 17-25.
1432   _____ . 1968. "Zum Papier der Leipziger Folio-Bibel von 1701."
        Gutenberg-Jahrbuch 43: 11-16.

1433 ————. 1970. "Papiermacherei 1587: Erstmals Erwähnung Einer Papiermühlebesichtigung in Einer Reisebeschreibung." *Marginalien* No. 38: 74.

1434 ————. 1970. "Schiller Als Papierliebhaber." *Marginalien* No. 40: 70-74.

1435 ————. 1963. "Die Stützerbacher Papiermühle im Roman." *Marginalien* No. 14 (December): 45-46.

1436 ————. 1953. *Thüringer Papiermühlen und Ihre Wasserzeichen.* Weimar, Germany: Thüringer Volksverlag.

1437 ————. 1961. "Unbeschriebene Blätter." *Marginalien* No. 12 (October): 21-25.

1438 ————. 1951. "Die Wasserzeichen-Papiere des Papierwerks Blankenburg im Wandel der Jahrhunderte." *Wochenblatt für Papierfabrikation* 79 (December 15): 771-777.

1439 Weisser, Erich. 1964. "Zur Geschichte der Papiermaschinen in Deutschland." *Papiergeschichte* 14 (July): 40-41.

1440 Wiedermann, Fritz. 1931. "Die Papiermühle in Reinerz." *Wochenblatt für Papierfabrikation* 62 (June 27): 619.

1441 ————. 1951. "Das Schicksal Alter Papiermühlen." *Papiergeschichte* 1 (December): 58.

1442 Zeltner, Eugen. 1913. "Gerber und Papierer in Freiburg i. Br. Bis Zum Ende des 16. Jahrhunderts." Unpublished Ph.D. dissertation, University of Freiburg.

1443 Zuman, F. 1948. "Přehlad Papíren na Moravě a ve Slezsku." *Papír a Celulósa* 4 (March-April): 15-17.

1444 ————. 1943. "Zlořády v Moravských a Slezských Papírnách v 18. Století (Příspěvek k Dějinám Papírnictvi na Moravé a ve Slzesku)." *Časopis Matice Moravské* 65: 85-111.

HUNGARY

1445 Bogdán, István. 1968. "Adalék a Pápai Papírmalom Történetéhez." *Papiripar* 12 (No. 4): 138-142.

1446 ————. 1962. "Egy 'Elégtelen' Papírtörténeti Kiadványról." *Papíripar és Magyar Grafika* 6 (January-February): 45-46.

1447 ————. 1959. "Az Első Magyar Papírmalok." *Papíripar és Magyar Grafika* 3 (March-April): 67-73.

1448 ————. 1974. "A Fehér Mív." *Papiripar* 18: 269-270.

1449 ————. 1960. "A Lékai (Hámori) Papírmalom a XVIII. Században." *Történelmi Szemle* 1: 46-93.

in Europe and the Soviet Union

1450 _____. 1963. *A Magyarországi Papiripar Története 1530-1900.* Budapest, Hungary: Akadémiai Kiadó.

1451 _____. 1963. "A Merített Papír a 14. Századtól a 19. Század Ekső Feléig." *Papíripar és Magyar Grafika* 7 (May-June): 110-115.

1452 _____. 1957. "XIX. Századi Vizjeleink." *Papíripar és Magyar Grafika* 1 (January-February): 23-31.

1453 _____. 1962. "A Papiralakok Változása a 14. Századtól a Szabványositásig." *Papiripar* 6: 237-246.

1454 _____. 1968. "Papírellátásunk és Papírkereskedelmünk a XVIII-XIX. Században." *Levéltári Közlemények* 39: 9-27.

1455 _____. 1973. "Papírfeldolgozó Iparunk a XIX. Században." *Papiripar* 17 (No. 1): 23-31.

1456 _____. 1973. "Papírfeldolgozó Iparunk a XIX. Században." *Papiripar* 17 (No. 4): 167-170.

1457 _____. 1956. "Papiripari Találmányok 1750-1830." *Papír- és Nyomdatechnika* 8: 125-129.

1458 _____. 1969. "Papírkereskedelem Budapesten a XIX. Században." *Papiripar* 13 (No. 5): 217-220.

1459 _____. 1965. "Papírkereskedelmünk a XIV-XVII. Században." *Századok* 99: 871-881.

1460 _____. 1956. "Papírkészítö Mesterek és Legények a XVIII. Században." *Papír és Nyomdatechnika* 8: 189-196.

1461 _____. 1960. "Papirmalom Alapitási Kisérletek Budán." *Papiripar* 4: 195-200.

1462 _____. 1953. "A Pécsi Papírmalom." *Papír- és Nyomdatechnika* 7 (February-March): 70-72.

1463 _____. 1962. "16.-17. Századi Papirkésitöink Életéböl." *Papiripar* 6: 128-131.

1464 _____. 1968. "Utánzott Vizjeleink a Századfordulón." *Papiripar* 12 (No. 1): 29-34.

1465 _____. 1970. "A Vízjel Mint Védjegy." *Papiripar* 14 (February): 64-70.

1466 _____. 1964. "A 'Vízjelirás' Fejlődése." *Papíripar és Magyar Grafika* 8 (September-October): 170-178.

1467 Dankanits, Ádám. 1959. "Adalék a Felenyedi Papírmalom Történetéhez." *Magyar Könyvszemle* 75 (July-September): 283-287.

1468 Eineder, Georg. 1960. *The Ancient Paper-Mills of the Former Austro-Hungarian Empire and Their Watermarks.* Hilversum, The Netherlands: The Paper Publications Society.

95

1469 Herepei, János. 1957. "Hol Volt a Gyulafehérvári Academicum Collegium Papirosmalma?" *Magyar Könyvszemle* 73 (October-December): 364-368.

1470 ————. 1961. "Újabb Megjegyzések a Felenyedi Papirosmalom Történetéhez." *Magyar Könyvszemle* 77 (August-October): 295-297.

1471 Jako, Zsigmond. 1961. Heltai Gáspár Papírmalma." *Magyar Könyvszemle* 77 (August-October): 290-295.

1472 Szendrey, István. 1963. "Adalék a XVIII. Századi Magyar Papírkészítés Történetéhez." *Magyar Könyvszemle* 79 (July-September): 240-242.

IRELAND

1473 Kelleher, Denis. 1950. "Papermaking in Ireland." *The Paper Maker* 19 (September): 21-29.

1474 ————, and Thomas P. O'Neill. 1960. "Early Paper Currency of the Bank of Ireland." *The Paper Maker* 29 (September): 1-11.

1475 Kennedy, Desmond. 1961. "The Hibernia Watermark." *The Paper Maker* 30 (February): 35-43.

1476 ————, and Alf MacLochlainn. 1960. "The Journals of the House of Commons: an Important Source of Irish Papermaking History." *The Paper Maker* 29 (February): 27-36.

1477 MacLochlainn, Alf. 1965. ". . . . Ever So Good or Ever So Cheap . . . ." *The Paper Maker* 34 (September): 19-27.

1478 O'Neill, Thomas P. 1962. "Irish Papermakers and Excise Duty on Paper, 1798-1861." *The Paper Maker* 31 (September): 33-38.

1479 Phillips, James W. 1958. "A Trial List of Irish Papermakers, 1690-1800." *The Library (Fifth Series)* 13 (March): 59-62.

ITALY

1480 Alibaux, Henry. 1939. "Papyrus und Papier auf Sizilien Mittelalter." *Wochenblatt für Papierfabrikation* 70 (January 7): 11-14.

1481 Beans, George Harry. 1938. *Some Sixteenth Century Watermarks Found in Maps Prevalent in the "Iato" Atlases.* Jenkintown, Pennsylvania: The George H. Beans Library.

1482 Bockwitz, Hans H. 1939. "Zur Frühzeit des Papiers auf Sizilien." *Philobiblon* 11 (March): 94-96.

1483 ————. 1939. "Neue Forschungen Zur Alt-Italienischen

Papiergeschichte." *Wochenblatt für Papierfabrikation* 70 (June 3): 469-473.

1484 _____ . 1942. "Papyrus und Papier auf Sizilien im Mittelalter." *Archiv für Buchgewerbe und Gebrauchsgraphik* 79: 148-150.

1485 _____ . 1938. " 'Pergamena Graeca'. Ein Papiergeschichtliches Problem." *Wochenblatt für Papierfabrikation* 69 (Special Edition): 6-9.

1486 Cecchetti, B. 1885. "Per la Storia Dell'Arte Della Carta Nelle Provincie Venete." *Archivio Veneto* 29: 419-420.

1487 Elliott, Harrison. 1952. "The Oldest Paper Mill in the Western World and Its Historical Background." *The Paper Maker* 21 (February): 39-46.

1488 Emery, Osvaldo. 1968. "Fabriano la Ville du Papier et du Peintre Gentile." *La Papeterie* 90 (December): 1302-1310.

1489 _____ . 1966. "Introducción del Papel en Europa." *Investigación y Técnica del Papel* 3 (April): 283-289.

1490 Faloci Pulignani, D.M. 1909. "Le Antiche Cartiere di Foligno." *Le Bibliofilia* 11 (June-July): 102-127.

1491 Fedrigoni, Antonio. 1966. *L'Industria Veneta Della Carta Dalla Seconda Dominazione Austriaca All'Unita d'Italia.* Torino, Italy: Ilte-Industria Libraria Tipografica Editrice.

1492 Gasparinetti, A.F. 1956. "Ein Altes Statut von Bologna Über die Herstellung und der Handel von Papier." *Papiergeschichte* 6 (July): 45-47.

1493 _____ . 1957. "Bartolo da Sassoferrato und Pietro Baldeschi." *Papiergeschichte* 7 (July): 50-52.

1494 _____ . 1956. "Frühe Papierherstellung in der Toskana." *Papiergeschichte* 6 (November): 68-70.

1495 _____ . 1966. "An Honor to Pace, Papermaker of Fabriano." *The Paper Maker* 35 (October): 31-34.

1496 _____ . 1954. "The Italian Papermen's Guild From Medieval Days to the Present." *The Paper Maker* 23 (September): 51-56.

1497 _____ . 1961. "Modena, ein Altes Zentrum der Papierherstellung in Italien." *Papiergeschichte* 11 (May): 35-36.

1498 _____ . 1960. "A Note About Paper Sales in XIV-Century Italy: Two Florentine Merchants Order Watermarked Paper From Fabriano in 1389." *The Paper Maker* 29 (February): 39-43.

1499 _____ . 1958. "Notes on Early Italian Papermaking." *The Paper Maker* 27 (February): 25-30.

1500 _____ . 1955. "Two Legendary Paper Mills." *The Paper Maker* 24 (February): 37-41.

1501 ———. 1952. "Über die Namen Alter Italienischer Papiermacher." *Papiergeschichte* 2 (April): 13-16.

1502 ———. 1956. "The Watermarked Names of XIV Century Italian Papermakers." *The Paper Maker* 25 (September): 15-26.

1503 Gerardy, T. 1968. "Einige Besonderheiten von Italienischen Papieren des 14. Jahrhunderts." *Papiergeschichte* 18 (December): 64-69.

1504 Irigoin, Jean. 1968. "La Datation des Papiers Italiens des XIIIe et XIVe Siècles." *Papiergeschichte* 18 (July): 49-52.

1505 ———. 1958. "Les Filigranes de Fabriano (Noms de Papetiers) dans les Manuscripts Grecs du Début du XIVe Siècle." *Scriptorium* 12: 44-50, 281-282.

1506 ———. 1959. "Filigranes Inédits de Fabriano (Début du XIVe Siècle)." *Papiergeschichte* 9 (October): 39-43.

1507 ———. 1960. "L'Introduction du Papier Italien en Espagne." *Papiergeschichte* 10 (July): 29-32.

1508 ———. 1963. "Les Origines de la Fabrication du.Papier en Italy." *Papiergeschichte* 13 (December): 62-67.

1509 Kotte, H. 1950. "Wie das Wasserzeichen in die Welt Gefommen Ist." *Allgemeine Papier-Rundschau* No. 3 (February 15): 90-93.

1510 Marchlewska, Jadwiga. 1971. "Wrażenia z Fabriano." *Przegląd Papierniczy* 27 (May): 186-188.

1511 Renker, Armin. 1952-1953. "Leonardo da Vinci und das Papier." *Imprimatur* 11: 197-198.

1512 ———. 1952. "Leonardo da Vinci und das Papier." *Papiergeschichte* 2 (November): 64.

1513 Ridolfi, Roberto. 1957. *Le Filigrane dei Paleotipi. Saggio Metodologico.* Firenze, Italy: Tipografia Giuntina.

1514 Rossini, Giuseppe. 1956. "Ulteriori Notizie Su la Cartiera. I Librai e le Prime Stampe Faentine." *Studi Romagnoli* 7: 283-292.

1515 Schlosser, Leonard B. 1976. "Some Early Milanese Paper Wrappers." In *A Pair on Paper: Two Essays on Paper History and Related Matters,* by Leonard B. Schlosser and Henry Morris. North Hills, Pennsylvania: Bird & Bull Press. 44-68.

1516 Valls i Subirà, Oriol. 1971. "Estudio Sobre la Trituración de los Trapos (1)." *Investigación y Técnica del Papel* 8 (April): 427-448.

1517 ———. 1970. "Estudio Sobre los Principios del Empleo de la Forma Con la Tela Fija, Seguido de Unas Notas Sobre los

Primeros Intercambios Papeleros Entre Italia y Cataluña (1)."
Investigación y Técnica del Papel 7 (April): 465-482.
1518   White, F.A. 1920. "Oldest Watermarks of Papal States and Nearby
Provinces." *Paper* 26 (June 30): 15-17.

LUXEMBURG

1519   Vekene, Emil van der. 1966. "Papiermühlen in Luxemburg 1689-
1919." *Gutenberg-Jahrbuch* 41: 13-17.

THE NETHERLANDS

1520   Barker, Ernest F. 1961. "270 Year Old Paper Mill Powered By Wind
and Gas." *Paper Trade Journal* 145 (October 16): 34-39.
1521   Bockwitz, Hans H. 1936. "Eine Alt-Holländische Papier-
Wassermühle." *Der Altenburger Papierer* 10: 552-554.
1522   _____. 1948. "Een Heksenverbranding in Het
Papiermakersbedrijf Aan Het Einde van de Zeventiende
Eeuw." *De Papierwereld* 3 (July): 9.
1523   Boer, P. 1955. "De Erste Papiermachine in Gelderland." *De
Papierwereld* 9 (June): 287-291.
1524   _____. 1954. "De Papiermolens te Egmond." *De Papierwereld* 8
(June): 266-270.
1525   _____, and H. Voorn. 1953. "Papierfabricage in de Nederlanden
in de Zestiende Eeuw, in Het Bijzonder te Dordrecht en te
Alkmaar." *De Papierwereld* 7 (February): 195-214.
1526   Bos, T.G.A. 1955. "Een Amsterdamsche Acte van 1573 Betreffende
een Papiermolen te Schophem Bij 's-Gravenvoeren (Prov.
Luik)." *Maasgouw LXIX* 74: 147-149.
1527   _____. 1957. "De Papiermolen(s) te Schophem Rond 1600."
Maasgouw LXXI 76: 79-82.
1528   Churchill, W.A. 1935. *Watermarks in Paper in Holland, England,
France, etc., in the XVII and XVIII Centuries and Their Interconnection.* Amsterdam, The Netherlands: M. Hertzberger &
Company.
1529   Enschedé, J.W. 1909. "Papier en Papierhandel in Noord-Nederland
Gedurende de Zeventiende Eeuw." *Tijdschrift Voor Boek- en
Bibliotheekwezen* 7: 97-111.
1530   _____. 1909. "Papier en Papierhandel in Noord-Nederland
Gedurende de Zeventiende Eeuw (II)." *Tijdschrift Voor Boek-
en Bibliotheekwezen* 7: 173-188.

1531   . 1909. "Papier en Papierhandel in Noord-Nederland Gedurende de Zeventiende Eeuw (III)." *Tijdschrift Voor Boek-en Bibliotheekwezen* 7: 205-211.

1532   Felix, D.A. 1952. "Een Gegateerd Handschrift op Papier uit de 9e Eeuw." *De Papierwereld* 6 (July): 398-400.

1533   Gerardy, T. 1969. "Von der Bluwen-Mühle I. Was Ist eine 'Bluwen' (Blöwi, Plauel, Blüli, Pleuel)?" *Papiergeschichte* 19 (June): 8-11.

1534   . 1958. "Die Hollandia-Wasserzeichen von F.C. Drewsen & Son 1822-35." *Papiergeschichte* 8 (February): 1-8.

1535   Herdies, H. 1960. "De Papiermolen te Vorst-Brüssel, Zijn Voorgaande Korremolen en Enkele Voormalige Toestanden van de Papiernijverheid in Brabant." *Eigen Schoon en de Brabander* 43: 224-235.

1536   Iongh, Jane de. 1934. *Van Gelder Zonen, 1784-1934*. Haarlem, The Netherlands: De Erven F. Bohn n.v.

1537   Kamminga, L. 1962. "De Papiermolen 'Het Springend Hart' Bij Makkum 1767-1906." *De Papierwereld* 17 (November): 115-118.

1538   Labarre, E.J. 1948. "An Interesting Find; an Historic Paper Maker's Mould." *World's Paper Trade Review* 130 (July 22): 215-216, 218, 220, 256.

1539   Langenbach, Alma. 1939. "Beziehungen der Westfälischen Papiermacherei zu den Niederlanden." *Gutenberg-Jahrbuch* 14: 47-51.

1540   . 1959. "Eine Papiermühle in Westfalischer Dichtung." *Papiergeschichte* 9 (November): 52-54.

1541   . 1958. "Die 'Rosenpapiermühle' zu Westig." *Papiergeschichte* 8 (September): 53-60.

1542   Marchlewska, Jadwiga. 1974. "Nieco Historü Papiernictwa Holandü." *Przegląd Papierniczy* 30 (April): 151-156.

1543   Pels, C. 1949. "De Eerste Papiermachine in Nederland." *De Papierwereld* 4: 24-26.

1544   Sabbe, Maurits. 1938. "Der Amsterdamer Papierhändler Vincent und die Antwerpener Drucker Moretus-Plantin." *Gutenberg-Jahrbuch* 13: 17-24.

1545   Tschudin, W.F. 1950. "Einige Bemerkungen Zur Reproduktion des Papiermacher-Kupferstichs von Jan und Kasper Luyken (Amsterdam 1694)." *Schweizerisches Gutenbergmuseum* 36: 21-23.

1546   Voorn, Henk. 1966. "Alkmaar en de Oudste Hollandse
Papiermolens." *De Papierwereld* 21 (March): 79-88, 91.

1547   ————. 1965. "Amsterdam en Angoulême." *De Papierwereld* 19
(June): 407-417.

1548   ————. 1955. "Die Anfänge der Papiermacherei in den
Niederlanden Bis Zum Ende des 16. Jahrhunderts."
Papiergeschichte 5 (May): 23-28.

1549   ————. 1967. "Anglo-Dutch Relations in Paper-Making History."
De Papierwereld 22 (October): 293-296.

1550   ————. 1955. "De Basterd- en Grauwpapiermolen 'De Veering'
te Koog Aan de Zaan." *De Papierwereld* 10 (December):
114-115.

1551   ————. 1956. "De Basterd-Wittepapiermolen 'De Veenboer' te
Zaandijk." *De Papierwereld* 11 (December): 106-107.

1552   ————. 1956. "De Basterdpapiermolen 'De Bok' te Wormerveer."
De Papierwereld 10 (January): 142.

1553   ————. 1956. "De Blauw- en Basterdpapiermolen 'De Oude
Voorn' te Wormerveer." *De Papierwereld* 10 (February): 158-
161, 168.

1554   ————. 1953. "De Blauwpapiermolen 'De Hobbezak' te Zaandijk
1670-1868." *De Papierwereld* 8 (October): 63-66.

1555   ————. 1956. "De Blauwpapiermolen 'De Vergulde Bijkorf' Alias
'De Bel' te Wormerveer." *De Papierwereld* 10 (March):
183-186.

1556   ————. 1954. "De Blauwpapiermolen 'Het Guiswijf' te Zaandijk
1670-1707." *De Papierwereld* 8 (March): 191-192.

1557   ————. 1955. "Drie Zaandamse Papiermolens." *De Papierwereld*
9 (July): 311-318.

1558   ————. 1957. "A Dutch Papermaker's Salute to American
Independence." *The Paper Maker* 26 (February): 7-9.

1559   ————. 1953. "Emperor of All Russia, Papermaker: the Little
Known Story of How Peter the Great Learned This Art at
Dutch Mills." *The Paper Maker* 22 (September): 33-36.

1560   ————. 1961. "Die Geschichte der Windpapiermühle 'De
Schoolmeester'." *Papiergeschichte* 11 (December): 84-87.

1561   ————. 1954. "De Grauwpapiermolen 'De Jonge Dolfijn' of
'Koperenberg' te Westzaan 1695-1825." *De Papierwereld* 9
(December): 128-132.

1562   ————. 1955. "De Grauwpapiermolen 'De Jonge Zwaan'
Bijgenaamd 'De Koger Oud' 1616-1855." *De Papierwereld* 10
(August): 7-10.

1563   ————. 1956. "De Grauwpapiermolen 'De Kaarsenmaker' te Koog Aan de Zaan." *De Papierwèreld* 10 (June): 260-261.

1564   ————. 1954. "De Grauwpapiermolen 'De Kauwer' of 'De Grauwe Papierbaal' te West-Zaandam 1616-1765." *De Papierwereld* 9 (August): 17-19.

1565   ————. 1955. "De Grauwpapiermolen 'De Koekoek' te Koog Aan de Zaan 1668-1847." *De Papierwereld* 9 (May): 266-269.

1566   ————. 1954. "De Grauwpapiermolen 'De Ruiter' te Westzaan 1695-1801." *De Papierwereld* 9 (October): 73-77.

1567   ————. 1954. "De Grauwpapiermolen 'De Salamander' te Koog Aan de Zaan 1660-1742." *De Papierwereld* 9 (November): 101-103.

1568   ————. 1956. "De Grauwpapiermolen 'De Soldaat' te Wormerveer." *De Papierwereld* 10 (January): 141-142.

1569   ————. 1954. "De Grauwpapiermolen 'De Vos' te West-Zaandam 1681-1735." *De Papierwereld* 8 (April): 210-212.

1570   ————. 1957. "Het Huis Assumburg te Westzaan." *De Papierwereld* 11 (May): 228-231.

1571   ————. 1960. "Das Hochdeutsche und Basler Papier im Holländischen Papierhandel." *Papiergeschichte* 10 (December): 77-80.

1572   ————. 1967. "Een Koninklijke Dichteres Bezingt Het Papier. Het Papiermakerslied van Carmen Sylva." *De Papierwereld* 22 (May-June): 141-144.

1573   ————. 1962. "The Last Wind-Driven Paper Mill in The Netherlands: the Schoolmaster." *The Paper Maker* 31 (March): 3-11.

1574   ————. 1969. "Martin Orges: de Papiermolen in Het Arnhernse Openluchtmuseum." *De Papierwereld* 24 (May): 127-133.

1575   ————. 1954. "Een Naamloze Grauwpapiermolen te Zaandink en de Papiermolen 'De Gans' 1605-1741." *De Papierwereld* 8 (July): 294-296, 307.

1576   ————. 1955. "Natural History in Watermarks." *Pulp and Paper Magazine of Canada* 56 (June): 93.

1577   ————. 1961. "De Oudste Hollandse Papiermolens." *De Papierwereld* 15 (July): 341-349.

1578   ————. 1951. "Papermaking and Papermakers of Long Ago in Holland." *The Paper Maker* 20 (September): 1-9.

1579   ————. 1950. "De Papiermaker van Porcelius." *De Papierwereld* 5 (August): 24-25.

1580   ————. 1956. "Een Papiermolen Bij Haarlem." *De Papierwereld* 11 (October): 61-63.

1581 ————. 1956. "De Papiermolen 'De Eendragt' te Wormer." *De Papierwereld* 10 (May): 238-242.

1582 ————. 1957. "De Papiermolen 'De Mol' en 'De Blauwe Papiermolen' te Krommenie." *De Papierwereld* 12 (August): 14-16.

1583 ————. 1957. "De Papiermolen 'De Veldkat'." *De Papierwereld* 11 (February): 159.

1584 ————. 1957. "Papiermolen 'De Witte Dolfijn' te Westzaan." *De Papierwereld* 11 (July): 285-286.

1585 ————. 1955. "De Papiermolen 'Het Guiskind' te Zaandijk." *De Papierwereld* 10 (November): 87-93, 97.

1586 ————. 1959. "Een Papiermolen in Drenthe." *De Papierwereld* 13 (July): 322a-322b.

1587 ————. 1962. "De Papiermolen van Louis de Geer." *De Papierwereld* 17 (December): 139-152.

1588 ————. 1955. "De Papiermolens der Familie Blauw. Deel I: de Voorgangers van Dirk Blauw." *De Papierwereld* 9 (February): 181-184.

1589 ————. 1955. "De Papiermolens der Familie Blauw. Deel II: de Directie van Dirk Blauw." *De Papierwereld* 9 (March): 209-213, 216-217.

1590 ————. 1955. "De Papiermolens der Familie Blauw. Deel III: de Erven Dirk Blauw." *De Papierwereld* 9 (April): 241-243.

1591 ————. 1973. *De Papiermolens in de Provincie Zuid-Holland.* Wormerveer, The Netherlands: Meijer.

1592 ————. 1953. "Die Papiermühle 'De Gooier' in Westzaan (Holland) 1670-1743." *Papiergeschichte* 3 (December): 81-84.

1593 ————. 1963. "Some Curious Early Experiments in Dutch Papermaking." *The Paper Maker* 32 (March): 13-19.

1594 ————. 1963. "Some Notes on the History of Dutch Paper Commerce." *The Paper Maker* 32 (September): 3-12.

1595 ————. 1951. "Twee Episoden Uit de Geschiedenis van de Houtslijp." *De Papierwereld* 6 (November): 123-128.

1596 ————. 1955. "De Vergulde Bijkorf Bijenaamd de Guisman 1662-1902." *De Papierwereld* 10 (September): 42-46.

1597 ————. 1955. "Van Westpennest tot Papierblad." *De Papierwereld* 10 (October): 66.

1598 ————. 1957. "De Westzaanse Grauwpapiermolen 'De Pronker'." *De Papierwereld* 11 (January): 130-132.

1599 ————. 1954. "De Westzaanse Grauwpapiermolen 'De Schoolmeester'." *De Papierwereld* 8 (February): 162-165, 168.

1600 ————. 1953. "De Witpapiermolen 'De Gooier' te Westzaan 1670-1743." *De Papierwereld* 7 (May): 309-312.

1601 ————. 1956. "De Witpapiermolen 'De Jonge Voorn' te Wormerveer." *De Papierwereld* 10 (April): 211-212.

1602 ————. 1953. "De Witpapiermolen 'De Kok' te Oost-Zaandam 1694-1735." *De Papierwereld* 8 (December): 109-112.

1603 ————. 1953. "De Witpapiermolen 'De Kruiskerk' te West-Zaandam 1712?-1841." *De Papierwereld* 8 (September): 37-41.

1604 ————. 1954. "De Witpapiermolen 'De Visser' te Zaandijk 1694-1796." *De Papierwereld* 8 (May): 237-240.

1605 ————. 1953. "De Witpapiermolen 'De Walrus' te West-Zaandam 1659-1749." *De Papierwereld* 7 (July): 356-359.

1606 ————. 1953. "De Witpapiermolen 'De Walvis' te West-Zaandam 1660-1817." *De Papierwereld* 8 (August): 15-17.

1607 ————. 1956. "De Witpapiermolens de Bonsem, de Wever en Het Fortuin en de Familie van der Ley." *De Papierwereld* 10 (July): 286-287.

1608 ————. 1956. "De Witpapiermolens de Bonsem, de Wever en Het Fortuin en de Familie van der Ley. Deel II." *De Papierwereld* 11 (August): 12-17.

1609 ————. 1965. "De Witpapiermolens de Bonsem, de Wever en Het Fortuin en de Familie van der Ley (Slot)." *De Papierwereld* 11 (September): 39-44.

1610 ————. 1956. "De Witte Papiermolen 'Het Herderskind' te Zaandijk." *De Papierwereld* 11 (November): 81-83.

1611 ————. 1957. "De Witte Veer te Zaandijk." *De Papierwereld* 11 (February): 158-159.

1612 ————. 1953. "De Zaandijker Grauwpapiermolen 'De Zemelzak'." *De Papierwereld* 7 (April): 277-278, 280.

1613 ————. 1954. "De Zaandijker Witpapiermolen 'De Herder' 1689-1842." *De Papierwereld* 8 (January): 131-134.

1614 ————. 1953. "De Zaandijker Witpapiermolens 'De Hoop' 1679-1849 en 'De Herderin' 1690-1830." *De Papierwereld* 8 (November): 79-86.

1615 Vries, Benjamin Willem de. 1957. *De Nederlandse Papiernijverheid in de Negentiende Eeuw.* The Hague, The Netherlands: Martinus Nijhoff.

1616 Weiss, Wisso. 1953. "Das Fortuna - Wasserzeichen." *Wochenblatt für Papierfabrikation* 81 (December 15): 901-906.

1617   Wescher, Hertha. 1949. "Papierlijming in Robert's Tijd." *De Papierwereld* 4: 18-19.

NORWAY

1618   Bull, Edvard. 1953. *Arbeidsfolk Forteller Fra Papirindustrien.* Oslo, Norway: Tiden Norsk Forlag.

1619   Fiskaa, Haakon M. 1939. "Om Anvendelse av Vannmerker i 17-1800-Årene For å Hindre Forfalskning av Pengesedler og Til Avsløring av Falske Dokumenter." *Papir-Journalen* 27 (July 17): 176-182.

1620   _____. 1939. "Om Anvendelse av Vannmerker i 17-1800-Årene For å Hindre Forfalskning av Pengesedler og Til Avsløring av Falske Dokumenter." *Papir-Journalen* 27 (August 15): 198-202.

1621   _____. 1934. "Eiker Papirmølle." *Papir-Journalen* 22 (June 5): 97-100.

1622   _____. 1934. "Fennefoss Papirmølle, 1804-1813. Et Glemt Anlegg." *Papir-Journalen* 22 (November 24): 223-226.

1623   _____. 1934. "Fennefoss Papirmølle, 1804-1813. Et Glemt Anlegg." *Papir-Journalen* 22 (November 30): 233-238.

1624   _____. 1936. "Gausa Papirmølle, 1849-1868." *Papir-Journalen* 24 (July 31): 149-152.

1625   _____. 1936. "Gausa Papirmølle, 1849-1868." *Papir-Journalen* 24 (August 15): 161-165.

1626   _____. 1936. "Gausa Papirmølle, 1849-1868." *Papir-Journalen* 24 (August 31): 174-178.

1627   _____. 1973. *Norske Papirmøller og Deres Vannmerker.* Oslo, Norway: Universitatsbiblioteket.

1628   _____. 1940. *Papiret og Papirhandelen i Norge i Eldre Tid.* Oslo, Norway: C.E. Petersen.

1629   _____. 1949. "Representative Norske Vannmerker 1695-1750." *Norsk Boktrykk-Kalender* n.v.: 17-41.

1630   _____. 1950. "Representative Norske Vannmerker 1750-1775." *Norsk Boktrykk-Kalender* n.v.: 46-67.

1631   _____. 1952. "Representative Norske Vannmerker 1775-1870." *Norsk Boktrykk-Kalender* n.v.: 19-33.

1632   _____. 1941. "Den Symbolske Betydning av de Eldste Vannmerker og Oprinnelsen Til Våre Papirnavn." *Papir-Journalen* 29 (January 31): 5-7.

1633 _____. 1941. "Den Symbolske Betydning av de Eldste Vannmerker og Oprinnelsen Til Våre Papirnavn." *Papir-Journalen* 29 (February 20): 16-18.

1634 _____. 1941. "Den Symbolske Betydning av de Eldste Vannmerker og Oprinnelsen Til Våre Papirnavn." *Papir-Journalen* 29 (February 28): 25-26.

1635 _____. 1941. "Den Symbolske Betydning av de Eldste Vannmerker og Oprinnelsen Til Våre Papirnavn." *Papir-Journalen* 29 (March 25): 34-37.

1636 _____. 1941. "Den Symbolske Betydning av de Eldste Vannmerker og Oprinnelsen Til Våre Papirnavn." *Papir-Journalen* 29 (March 31): 46-48.

1637 Mohn, Victor I. 1938. "Bentse Brug. En Gammel Papirfabrikks Saga." *Papir-Journalen* 26 (July 20): 159-161.

1638 _____. 1938. "Bentse Brug. En Gammel Papirfabrikks Saga." *Papir-Journalen* 26 (July 30): 171-173.

1639 Møller-Nicolaisen, N.A. 1938. "Forsøg Til Rekonstruktion af Tycho Brahes Papirmølle." *Nordisk Astronomisk Tidsskrift* 19: 42-60.

1640 _____. 1954. "Tycho Brahe und Seine Papiermühle." *Papiergeschichte* 4 (November): 57-67.

1641 _____. 1930. "Et Tycho Brahe-Minde Paa Hven." *Nordisk Astronomisk Tidsskrift* 11: 122-128.

1642 _____. 1933. "Tycho Brahes Papirmølle." *Nordisk Astronomisk Tiddskrift* 14: 85-95.

1643 _____. 1934. "Tycho Brahes Papirmølle." *Nordisk Astronomisk Tidsskrift* 15: 121-128.

1644 Nordstrand, Ove K. 1958-1959. "Vandmaerker Fra Frederik II's og Tycho Brahes Papiermöller." *Fund og Forskning i Det Kongelige Biblioteks Samlinger* 5-6: 218-221.

1645 Voorn, Henk. 1955. "Bentse Bruk: the First Paper Mill in Norway." *The Paper Maker* 24 (September): 31-41.

1646 _____. 1955. "De Eerste Noorse Papiermakers Kwamen Uit Zaandam." *De Papierwereld* 9 (January): 157-162.

1647 _____. 1956. "Old Norwegian Paper Mills: More About Bentse Bruk." *The Paper Maker* 25 (February): 1-8.

1648 _____. 1959. *The Paper Mills of Denmark & Norway and Their Watermarks.* Hilversum, The Netherlands: The Paper Publications Society.

POLAND

1649   Budka, Włodzimierz. 1928. "Filigrany z Herbami Łodzia i Lis."
        Silva Rerum 4: 180.
1650   _____. 1964. "Die Papierfabrik des Zamoyski-Majorats in
        Hamernia." *Papiergeschichte* 14 (December): 63-65.
1651   _____. 1970. "Papiernia Biskupów Krakowskich w Mędrowie."
        Przegląd Papierniczy 26 (September): 314-315.
1652   _____. 1928. "Papiernia Przysuska w 1777 Roku." *Przegląd
        Bibljoteczny* 2 (January-March): 53-54.
1653   _____. 1935. "Papiernia w Balicach." *Archeion* 13: 30-49.
1654   _____. 1929. "Papiernia w Kościelcu Śląskim." *Przegląd
        Bibljoteczny* 3: 30-34.
1655   _____. 1931. "Papiernia w Odrzykoniu i Mniszku." *Przegląd
        Bibljoteczny* 5 (January-March): 61-66.
1656   _____. 1952. "Papiernia w Sukowie." *Przegląd Papierniczy* 8:
        160-164.
1657   _____. 1954. "Papiernie Poznańskie." *Przegląd Papierniczy* 10
        (July): 216-221.
1658   _____. 1954. "Papiernie Poznańskie." *Przegląd Papierniczy* 10
        (August): 251-253.
1659   _____. 1948. "Papiernie w Królestwie Polskim w r. 1858."
        Przegląd Papierniczy 4 (May): 97-98.
1660   _____. 1953. "Papiernie w Księstwie Warszawskim i Krolestwie
        Polskim (1810-1830)." *Przegląd Papierniczy* 9 (April):
        121-123.
1661   _____. 1956. "Papiernie w Lublinie i Kocku." *Archeion* 25:
        257-275.
1662   _____. 1970. "Papiernie w Mostkach (Zimnowodzie), Drugni i
        Słopcu." *Przegląd Papierniczy* 26 (August): 278-280.
1663   _____. 1929. "Papiernie w Nowym Stawie i Radomyślu."
        Przegląd Bibljoteczny 3: 520-524.
1664   _____, editor. 1971. *Papiernie w Polsce XVI Wieku*. Warsaw,
        Poland: Ossolineum.
1665   _____. 1928. "Urądzenie Papierni Polskiej w Początku XIX
        Wieku." *Silva Rerum* 4: 30.
1666   _____. 1956. "Wykaz Fabryk Papieru w Królestwie Polskim z r.
        1823." *Przegląd Papierniczy* 12 (August): 248-252, XXXI.
1667   _____. 1956. "Załozenie Papierni w Kiszewie." *Przegląd
        Papierniczy* 12 (April): 115-117.

1668 _____ , and J. Siniarska-Czaplicka. 1965. "Papiernia Arcybiskupów Gnieźnieńskich w Kęszycach Pod Łowiczem." *Archeion* 42: 229.

1669 Ciesielski, Augustyn. 1969. "Papiernia Nad Kamiennym Potokiem Koło Sopotu." *Przegląd Papierniczy* 25 (August): 282-285.

1670 _____ . 1967. "Papiernia w Mniszku i Jej Papiernicy." *Przegląd Papierniczy* 23 (August): 277-279.

1671 Decker, Viliam. 1957. "Dzieje Papierni w Popradzie." *Przegląd Papierniczy* 13 (June): 181-184.

1672 _____ . 1956. "Papiernia w Wielkiej (Velkej)." *Przegląd Papierniczy* 12 (November): 341-343.

1673 Gajerski, Stanisław Franciszek. 1972. "Papiernia Szkielska Oraz Jej Poddani w Roku 1765." *Przegląd Papierniczy* 28 (January): 32-34.

1674 _____ . 1974. "Papiernia Szkielska w Roku 1771." *Przegląd Papierniczy* 30 (June): 230-232.

1675 Gębarowicz, Mieczysław. 1966. "Z Dziejów Papiernictwa XVI-XVIII W." *Roczniki Biblioteczne* 10 (Nos. 1-2): 1-84.

1676 Kamykowski, L. 1935. "Papiernia Lubelska." *Pamiętnik Lubelski* 2: 160.

1677 Kędra, Edward. 1974. "Herb Torunia w Filigranach Papierni Ziemi Chełmińskiej." *Przegląd Papierniczy* 30 (February): 72-74.

1678 Kuczyński, Stefan Krzysztof. 1968. "Jeszcze o 'Patriotycznym' Znaku Wodnym z Papierni w Soplu." *Przegląd Papierniczy* 24 (May): 174-175.

1679 _____ . 1964. " 'Patriotyczny' Znak Wodny z Papierni Sopelskiej." *Przegląd Papierniczy* 20 (March): 93.

1680 Laucevičius, E. 1967. *Popierius Lietuvojc XV-XVIIIa.* Vilnius, Lithuania SSR: Publishing House 'Mintis'. 2 Volumes.

1681 Łowmiański, Henryk. 1924. "Papiernie Wileńskie XVI Wieku." *Ateneum Wileńskie* 2: 409-422.

1682 Madyda, W. 1950. "Postnowienia Rzemiosla Papierniczego z r. 1546." *Przegląd Papierniczy* 6 (January): 33-34.

1683 Maleczyńska, K. 1961. *Dzieje Starego Papiernictwa Śląskiego.* Warsaw, Poland: Zakład Narodowy im. Ossolińskich.

1684 Marchlewska, Jadwiga. 1972. "Z Dziejów Nowszych i Najnowszych Dusznik." *Przegląd Papierniczy* 28 (July): 233-237.

1685 _____ . 1963. "O Papierze w Karnawale." *Przegląd Papierniczy* 19 (January): 25, 28.

1686 _____ . 1970. "A Short Outline of Polish Papermaking History From 1491 to 1945." *The Paper Maker* 39 (September): 24-37.

1687 Nawrocki, Stanisław. 1957. "Kilka Uwag o Papierni Tzw. Oleśnickiej Czyli Chodzieskiej." *Przegląd Papierniczy* 13 (February): 59-60.

1688 _____. 1955. "Papiernie w Bledzianowie i Olszynie w Drugiej Połowie XVIII Wieku." *Przegląd Papierniczy* 11 (April): 118-119.

1689 _____. 1958. "Papiernie w Powiecie Ostrzeszowskim w XIX Wieku." *Przegląd Papierniczy* 14 (December): 374-376.

1690 Pabich, Franciszek. 1965. "Autor Nieśmiertelnego Hymnu Narodowego, Józef Wybicki - Budowniczym Papierni w Będominie w Pow.koscierskim." *Przegląd Papierniczy* 21 (July): 218-220.

1691 _____. 1964. "Dzieje Papierni Chojnickiej." *Przegląd Papierniczy* 20 (August): XXXI-XXXII.

1692 _____. 1974. "Z Dziejów Papierni Słupskiej." *Przegląd Papierniczy* 30 (May): 197-199.

1693 _____. 1968. "Kilka Kart z Dziejów Papierni w Lubiczu Nad Drwęcą." *Przegląd Papierniczy* 24 (April): 138-140.

1694 _____. 1963. "Moje Osiagnięcia w Zakresie Poszukiwania i Selekcjonowania Znaków Wodnych Papierni Pomorza i Kaszub." *Przegląd Papierniczy* 19 (January): 23-24.

1695 _____. 1970. "Papiernia Nad Potokiem Karlikowskim." *Przegląd Papierniczy* 26 (March): 104-105.

1696 _____. 1973. "Papiernie Nadnoteckie." *Przegląd Papierniczy* 29 (August): 290-294.

1697 Pencak, T. 1958. "Papiernia Ordynacji Zamojskiej w Hamerni." *Archeion* 28: 159-177.

1698 Piekosinski, F., J. Ptasnik, and K. Piekarski. 1971. *Papiternie w Polsce XVI Wieku*. Warsaw, Poland: Ossolineum.

1699 Ptasnik, Joannes. 1953. "Frühe Papiermacherei in Polen." *Papiergeschichte* 3 (November): 62-69.

1700 Rączka, Zofia. 1966. "Początki Przemysłu Papierniczego w Zywiecczyźnie." *Przegląd Papierniczy* 22 (February): 60-63.

1701 Rożycki, Edward. 1975. "Z Dziejów Kartownictwa XVI-XVII Wieku." *Przegląd Papierniczy* 31 (January): 31-33.

1702 Sarnecki, Kazimierz. 1959. "Czy Nad Papiernią na Prądniku Wielkim Unosił Sie Dym?" *Przegląd Papierniczy* 15 (January): 25-26.

1703 _____. 1974. "Czy Przed Pięciu Wiekami Robiono w Gdańsku Papier?" *Przegląd Papierniczy* 30 (June): 228-230.

1704 _____. 1954. "Dokumenty Dolyczące Załozenia Pierwszej Papierni w Warszawie." *Przegląd Papierniczy* 10 (April): 119-121.

1705 _____. 1957. "Z Dziejów Papiernictwa Warszawskiego." *Przegląd Papierniczy* 13 (July): 220-221.

1706 _____. 1958. "Z Historii Papiernictwa w Dawnej Polsce." *Kwartalnik Historii Nauki i Techniki* 3: 223-241.

1707 _____. 1959. "Karol Chrystian Langsdorf (1757-1834) Wykładowca Technologii Papiernictwa i Autor Podręcznika o Wyrobie Papieru." *Przegląd Papierniczy* 15 (September): 283-286.

1708 _____. 1963. "Papier Warszawskich Druków Elerta i Spadkobierców z Lat 1643-1678." *Przegląd Papierniczy* 19 (May): 162-164.

1709 _____. 1954. "Papiernia w Bolimowie." *Przegląd Papierniczy* 10 (August): 253-254.

1710 _____. 1964. "Papiernia w Dusznikach." *Przegląd Papierniczy* 20 (December): 407-408.

1711 _____. 1970. "Skąd Pochodzi Slowo 'Arkusz'." *Przegląd Papierniczy* 26 (March): 105.

1712 _____. 1964. "Zaopatrzenie w Papier Warszawskiej Drukarni Jana Rossowskiego z Lat 1624-1634." *Przegląd Papierniczy* 20 (July): 230-231.

1713 _____. 1965. "Znaki Wodne Druków Warszawskich Jana Trełpińskiego z Lat 1636-1647 i Karola Fryderyka Schreibera z Lat 1674-1691." *Przegląd Papierniczy* 21 (April): 118-119.

1714 Siniarska-Czaplicka, Jadwiga. 1959. "Z Dziejów Papierni w Pabianicach." *Przegląd Papierniczy* 15 (June): 179-183.

1715 _____. 1969. *Filigrany Papierni Położonych Na Obszarze Rzeczypospolitej Polskiej Od Początku XVI do Połowy XVIII Wieku.* Warsaw, Poland: Ossolineum.

1716 _____. 1958. "Historia Papierni w Jeziornie do r. 1939." *Przegląd Papierniczy* 14 (June): 186-190.

1717 _____. 1958. "Historia Papierni w Jeziornie do r. 1939." *Przegląd Papierniczy* 14 (July): 208-212.

1718 _____. 1963. "Papiernia w Kopcu Koło Białej." *Przegląd Papierniczy* 19 (September): 296-297.

1719 _____. 1966. "Papiernictwo na Ziemiach Środkowej Polski w 1. 1750-1850." *Studia z Dziejów Rzemiosła i Przemysłu* 6: 150-154.

1720   ————. 1967. "Papiernicy Angielscy w Czerpalniach Królestwa Polskiego." *Przegląd Papierniczy* 23 (November-December): 402-405.

1721   ————. 1969. "Papiernie Nad Czarną Staszowską." *Przegląd Papierniczy* 25 (November): 383-386.

1722   ————. 1967. "Papiernie Olwodu Opoczyńskiego." *Przegląd Papierniczy* 20 (December): 408-411.

1723   ————. 1959. "Próba Wytypowania Metody Zbierania i Klasyfikowania Polskich Znaków Wodnych na Przykładzie Papierów Młynów Mazowsza w Latach 1750-1850." *Przegląd Papierniczy* 15 (July): 215-221.

1724   ————. 1967. "Przywileje i Obowiązki Papierników w Dawnej Polsce." *Przegląd Papierniczy* 23 (February): 67-70.

1725   ————. 1975. "Warszawskie Fabryki Obić Papierowych." *Przegląd Papierniczy* 31 (February): 77-78.

1726   ————. 1971. "Wyrób Polskich Kart do Gry w Latach 1500-1650." *Przegląd Papierniczy* 27 (June): 219-220.

1727   Sobalski, Franciszek. 1969. "Dzieje Częstochowskiej Papierni (do Roku 1914)." *Przegląd Papierniczy* 25 (April): 131-133.

1728   ————. 1964. "Z Dziejów Fabryki Obić Papierowych w Częstochowie." *Przegląd Papierniczy* 20 (May): 166-167.

1729   ————. 1963. "Papiernia w Łojkach Koło Częstochowy." *Przegląd Papierniczy* 19 (October): 331.

1730   ————. 1968. "Przemysł Papierniczy w Królestwie Polskim na Przełomie XIX i XX Wieku." *Przegląd Papierniczy* 24 (November): 396-401.

1731   Szubert, Aleksandra. 1974. "Papiernie w Okręgu Lwowskim w Latach 1854-1870, na Podstawie Sprawozdań Izly Handlowo-Przemysłowej we Lwowie." *Przegląd Papierniczy* 30 (October): 401-404.

1732   Tomaszewska, Wanda. 1966. "Z Dziejów Zabytkowej Papierni w Dusznikach." *Przegląd Papierniczy* 22 (May): 168-173.

1733   ————. 1959. "Historia Zabytkowej Papierni w Dusznikach." *Przegląd Papierniczy* 15 (November-December): 358-359.

1734   Walczy, S. 1956. "Rekonstrukcja Zabudowań Papierni Lubelskiej Według Widoku Brauna." *Archeion* 25: 276-280.

1735   Wawrzeńczak, Andrzej. 1974. "Papiery Archiwum Rodzinnego Bohdana Marconiego." *Przegląd Papierniczy* 30 (July): 282-284.

1736   Wegner, J. 1953. "Instruktarz Ekonomiczny Dóbr Nieborowskich w Roku 1777." *Teki Archiwalne* 1: 9-43.

1737   Winczakiewicz, Andrzej. 1965. "Franciszek Jeziorański." *Przegląd Papierniczy* 21 (October): 322-324.

1738   _____. 1974. "Wielkie Łosiny." *Przegląd Papierniczy* 30 (October): 400-401.

1739   Wyrwa, Juliusz. 1963. "130 Lat Zywieckiej Fabryki Papieru." *Przegląd Papierniczy* 19 (December): 374-375.

1740   Zuman, F. 1921. "Privilegium Papírny Jáchymovské." *Památky Archeologické* 32: 259-260.

1741   Żurowski, Stanisław. 1958. "Estimation of Production Capacity of Polish Paper Mills in the XV Century." *Bibliographical Annals* 2: 399-429.

1742   _____. 1956. "Z Historii Techniki Wyrobu Papieru w Poznaniu." *Przegląd Papierniczy* 12 (January): 12, 21-23.

1743   _____. 1955. "Jeszcze o Michale Eldsnerze - Szesnasłowiecznym Papierńiku Poznańskim." *Przegląd Papierniczy* 11 (June): 185-186.

1744   _____. 1955. "Jeszcze o Michale Eldsnerze - Szesnasłowiecznym Papierńiku Poznańskim." *Przegląd Papierniczy* 11 (July): 217-220.

1745   _____. 1955. "Jeszcze o Michale Eldsnerze - Szesnasłowiecznym Papierńiku Poznańskim." *Przegląd Papierniczy* 11 (September): 281-283.

1746   _____. 1958. "Obliczanie Wysokości Produkcji XVI-Wiecznych Papierni Polskich." *Roczniki Biblioteczne* 2 (Nos. 3-4): 399-424.

1747   _____. 1962. "Przyszły Polski Ośrodek Badań Nad Historia Papieru." *Roczniki Biblioteczne* 6 (Nos. 3-4): 135-145.

PORTUGAL

1748   Voorn, Henk. 1961. "Early Papermaking in Portugal." *The Paper Maker* 30 (September): 15-27.

RUMANIA

1749   Blücher, Gebhard. 1967. "Filigranele Brasovene Si Tipăuturile Chirilice Din Secolul al XVI-Lea." *Revista Bibliotecilor* 20: 421-426.

SCOTLAND

1750   Thomson, Alistar G. 1974. *The Paper Industry in Scotland, 1590-1861*. New York: Abner Schram.

1751   _____. 1965. "The Paper Industry in Scotland 1700-1861." Unpublished Ph.D. dissertation, University of Edinburgh.

1752   Waterston, R. 1945. "Early Paper Making Near Edinburgh." *The Book of the Old Edinburgh Club* 25: 55-60.

1753   _____. 1949. "Further Notes on Early Paper Making Near Edinburgh." *The Book of the Old Edinburgh Club* 27: 44-47.

SOVIET UNION

1754   Abramyan, A. 1941. "Pewaya Bumazhnaya Fabrika v Armenii." *Bumazhnaya Promîshlennost* 'No. 5: 49-52.

1755   Burian, V. 1954. "Filigrány Olomoucké Papirny v 16. Stol." *Sborni Sbornik SLUKO-B* 11: 123-124.

1756   Chylík, J. 1958. "Papirnici na Moravě Kdysi." *Vlastivědný Věstnik Moravský* 13: 229-231.

1757   Geraklitov, Aleksnadr Aleksandrovich. 1963. *Filigrani XVIII Veka na Bumage Rukopisnyh i Pečatnyh Dokumentov Russkogo Proishoždenija*. Moscow, Soviet Union: Isdvo Academy Hayk.

1758   Jenšs, J. 1960. "Papīrs un tā Ūdensyzīmes Latvijā XIX gs. Pirmajā Pusē." *Vestures Problēmas* 3: 175-236.

1759   Klepikov, Sokrat Aleksandrovich. 1952. "Filigrani i Shtempeli Bumag Russkogo Proizvodstva XVIII-XX vv." *Otdela Rukopiseĭ Gos. Biblioteki im V.I. Lenina* 13: 57-122.

1760   _____. 1959. *Filigrani i Shtempeli na Bumage Russkogo i Inostrannogo Proizvodstva XVII-XX vv.* Moscow, Soviet Union.

1761   _____. 1968. "Ispol'zovanie Filigranej v Rabote s Nedatirovannymi Rukopisjami i Pečatnymi Knigami XIII-XVI Vikov." *Sovetskie Arhivy* No. 6: 50-57.

1762   _____. 1963. "Russian Watermarks and Embossed Paper-Stamps of the Eighteenth and Nineteenth Centuries." *Papers of the Bibliographical Society of America* 57 (Second Quarter): 121-128.

1763   Kosharnivs' kii, M. 1930. "Z Istorii Staroi Paperovoi Promislovosti na Chernihivshchini." *Bibliolohichni Visti* No. 24: 5-43.

1764   _____. 1930. "Z Istorii Staroi Paperovoi Promislovosti na Chernihivshchini." *Bibliolohichni Visti* No. 25: 49-70.

1765   Krip´yakenich, I. 1926. "Naĭdavnishi Papirni na Ukraini."
        Bibliolohichni Visti No. 10: 64-65.
1766   Kukushkina, M.V. 1958. "Filigrani na Bumage Russkikh Fabrik
        XVIII-Nachala XIX v." *Istoricheskiĭ Ocherk i Obzor Fondov
        Rukopisnogo Otdela Biblioteki Akademii Nauk SSSR* 2:
        285-371.
1767   Laucevičius, E. 1967. *Popierius Lietuvoje XV-XVIIa.* Vilnius,
        Lithuania SSR: Publishing House 'Mintis'. 2 Volumes.
1768   _____. 1961. "Senoji Prienu Popieriaus Dirbtuve." *Svyturys* No.
        16 (August): 14.
1769   Likhachĕv, N.P. 1891. "Bumaga i Drevneĭshie Bumazhnye Mel´nitsy
        v Moskovskom Gosudarstve." *Zapiski Imp. Russkogo
        Arkheologicheskogo Obshchestva* 5: 237-342.
1770   Matsyuk, O.Y. 1974. *Papir ta Filihrani na Ukrayins´kykh Zemlyakh
        (XVI-Pochatok XX st.).* Kiev, Soviet Union: Naukova Dumka.
1771   Romanovs´kiĭ, V. 1926. "Do Istorii Papirnitstva na Ukraini."
        Bibliolohichni Visti No. 11: 73-74.
1772   Sičynśkyj, V. 1941. "Papierfabriken in der Ukraine im XVI.-XVIII.
        Jahrhundert." *Gutenberg-Jahrbuch* 16: 23-29.
1773   Šilhan, J. 1956. "K Dějinám no Jstarštch Papirnen v Jihlavském
        Kraji." *Vlastivědný Sborník Vysočiny* 1: 118-143.
1774   _____. 1962. "Špitálská Papírna na Starém Brně." *Brno v
        Minulosti a Dres* 4: 70-84.
1775   Stepanov, A. 1928. "Bumazhnaya Fabrika Sheremetevskikh
        Krepostnîkh Toropovikh." *Zapiski Istoriko-Bîtovogo Otdela,
        Gosudarstvennîi Ruskiĭ Mezeĭ* 1: 253-256.
1776   Uchastkina, Zoya Vasil´Evna. 1961. "Characteristics of Watermarks
        in Russian Handmade Paper." *The Paper-Maker (London)* 142
        (September): 61-65.
1777   _____. 1961. "Eigenümlichkeiten der Wasserzeichen in
        Russischen Handgeschöpften Papieren." *Papiergeschichte* 11
        (December): 89-93.
1778   _____. 1962. *A History of Russian Hand Paper-Mills and Their
        Watermarks.* Hilversum, The Netherlands: The Paper Publica-
        tions Society.
1779   _____. 1972. *Razvitie Bumazhnogo Proizvodstva v Rossii.*
        Moscow, Soviet Union: 'Lesnaya Promyshlennost'.
1780   _____. 1956. "Vodyanîe Znaki Russkoi Bumagi." *Trudî
        Instituta Istorii Estestvoznaniya i Tekhniki Akademii Nauk
        SSSR* 12: 312-337.

1781 _____. 1959. "Die Wasserzeichen des Russischen Papiers (III)." *Zellstoff und Papier* 8 (January): 26-29.

1782 Zaozerskaya, E.I. 1928. "Bogoroditskiĭ Bumazhniĭ Zavod Pervoĭ Polovinî XVIII v." *Trudî Gos. Istoricheskogo Muzeya* 4: 163-179.

SPAIN

1783 Asenjo Martínez, José Luis. 1970. "La Calidad del Papel Catalán a Fines del Siglo XVIII." *Investigación y Técnica del Papel* 7 (July): 653-670.

1784 _____. 1968. "Evolución de la Localización Provincial Papelera en España." *Investigación y Técnica del Papel* 5 (July): 615-630.

1785 _____. 1966. "Las Filigranas de Papel de 'Arrigorriaga' Hasta 1936." *Investigación y Técnica del Papel* 3 (July): 849-889.

1786 _____. 1966. "La Importación en España de Papel Contínuo en 1841." *Investigación y Técnica del Papel* 3 (January): 9-16.

1787 _____. 1970. "Las Ordenanzas Papeleras de 1777." *Investigación y Técnica del Papel* 7 (January): 31-45.

1788 _____. 1968. "Volaración de Muestras Catalanas de Papel a Mediados del Siglo XVIII." *Investigación y Técnica del Papel* 5 (October): 901-914.

1789 Basanta Campos, José Luis. 1966. "Algunas Adiciones a la Historia de la Fabricación del Papel en Galicia del Siglo XVIII a Nuestros Días." *Investigación y Técnica del Papel* 3 (January): 23-41.

1790 _____. 1967. "Filigranas en Documentos Gallegos. Archiv Histórico Provincial de Pontevedra." *Investigación y Técnica del Papel* 4 (October): 877-893.

1791 _____. 1968. "Las Marcas de Agua en los Incunables Gallegos." *Investigación y Técnica del Papel* 5 (October): 921-937.

1792 Bielza de Ory, Vicente. 1973. "Los Focos Españoles de Producción Papelera en el Pasado: Factores de Localización." *Investigación y Técnica del Papel* 10 (April): 387-413.

1793 Emery, Osvaldo. 1966. "Introducción del Papel en Europa." *Investigación y Técnica del Papel* 3 (April): 283-289.

1794 Gayoso Carreira, Gonzalo. 1973. "Antigua Nomenclatura Papelera Española." *Investigación y Técnica del Papel* 10 (January): 29-53.

1795 _____. 1970. "Apuntes Para la Historia del Papel en Toledo, Ciudad Real y el Antiguo Reino de Murcia." *Investigación y Técnica del Papel* 7 (April): 443-456.

1796 _____. 1971. "Apuntes Para la Historia Papelera de Andalucía." *Investigación y Técnica del Papel* 8 (October): 1081-1103.

1797 _____. 1972. "Apuntes Para la Historia Papelera de Cataluña." *Investigación y Técnica del Papel* 9 (April): 411-422.

1798 _____. 1970. "Apuntes Para la Historia Papelera del Antiguo Reino de Valencia." *Investigación y Técnica del Papel* 7 (October): 1059-1091.

1799 _____. 1972. "Características del Papel del 'Breviario Mozárabe' de Silos." *Investigación y Técnica del Papel* 9 (January): 85-96.

1800 _____. 1966. "Dos Españoles, los Gallegos Antonio y Miguel, Introdujeron la Fabricación del Papel en Germania y Basilea." *Investigación y Técnica del Papel* 3 (July): 589-611.

1801 _____. 1965. "La Fabricación del Papel en Galicia del Siglo XVIII a Nuestros Días." *Investigación y Técnica del Papel* 2 (April): 193-223.

1802 _____. 1969. "La Filigrana de la 'Mano' en Documentos de Galicia (España), de los Siglos XVI y XVII." *Investigación y Técnica del Papel* 6 (October): 1069-1086.

1803 _____. 1965. "Galicia y el Papel, del Siglo XI al de la Ilustración." *Investigación y Técnica del Papel* 2 (January): 55-66.

1804 _____. 1969. "Historia Papelera de Aragón." *Investigación y Técnica del Papel* 6 (April): 425-442.

1805 _____. 1968. "Historia Papelera de la Provincia de Avila." *Investigación y Técnica del Papel* 5 (January): 73-84.

1806 _____. 1971. "Historia Papelera de la Provincia de Castellón de la Plana." *Investigación y Técnica del Papel* 8 (April): 411-426.

1807 _____. 1967. "Historia Papelera de la Provincia de Cuenca." *Investigación y Técnica del Papel* 4 (April): 349-366.

1808 _____. 1968. "Historia Papelera de la Provincia de Guadalajara." *Investigación y Técnica del Papel* 5 (October): 979-1007.

1809 _____. 1968. "Historia Papelera de la Provincia de Segovia." *Investigación y Técnica del Papel* 5 (April): 359-374.

1810 _____. 1971. "Historia Papelera de la Provincia de Valencia." *Investigación y Técnica del Papel* 8 (July): 697-740.

1811 _____. 1968. "Historia Papelera de la Provincia de Valladolid." *Investigación y Técnica del Papel* 5 (July): 631-649.

1812 ————. 1967. "Historia Papelera de los Provincias de Burgos, Logroño, Soria y Santander." *Investigación y Técnica del Papel* 4 (October): 931-946.

1813 ————. 1970. "Historia Papelera de los Provincias de Guipúzcoa y Alava." *Investigación y Técnica del Papel* 7 (January): 75-105.

1814 ————. 1969. "Historia Papelera de los Provincias de Vizcaya y Navarra." *Investigación y Técnica del Papel* 6 (July): 775-791.

1815 ————. 1967. "Historia Papelera de los Regiones Leonesa, Ertremeña y Principado de Asturias." *Investigación y Técnica del Papel* 4 (July): 613-622.

1816 ————. 1969. "Historia Papelera de Madrid y Su Provincia." *Investigación y Técnica del Papel* 6 (January): 49-84.

1817 ————. 1973. "Interesantes Filigranas de un Pliego de Papel." *Investigación y Técnica del Papel* 10 (April): 459-465.

1818 ————. 1972. "Notas Sobre una Pequeña Colección de Filigranas." *Investigación y Técnica del Papel* 9 (October): 1015-1029.

1819 ————. 1966. "El Primer Molino Papelero de Guipúzcoa." *Investigación y Técnica del Papel* 3 (April): 307-312.

1820 Irigoin, Jean. 1960. "L'Introduction du Papier Italien en Espagne." *Papiergeschichte* 10 (July): 29-32.

1821 Labayen, Antonio María. 1970. "En un Lugar Llamado Erasote." *Investigación y Técnica del Papel* 7 (October): 1117-1124.

1822 Mena, Ramon. 1926. *Filigranas: O Marcas Transparentes en Papeles de Nueva España, del Siglo XVI.* Mexico: Monografias Bibliograficas Mexicanas.

1823 Real, José Sánchez. 1973. "Filigranas por Describir." *Investigación y Técnica del Papel* 10 (April): 493-513.

1824 Schulte, Ulman. 1961. "Beitrag zur Spanischen Papiergeschichte." *Papiergeschichte* 11 (May): 37-44.

1825 ————. 1963. "Besondere Katalanische Schöpfformen." *Papiergeschichte* 13 (October): 30-35.

1826 ————. 1962. "Einige Bemerkungen zu den Zick-Zack-Linien in Frühspanischen Papieren (Korreferat)." *Papiergeschichte* 12 (February): 7-9.

1827 Spahr, Wilhelm. 1952. "Juan Gabriel Romani und Seine Nachkommen: ein Kapitel Spanischen Papiergeschichte." *Wochenblatt für Papierfabrikation* 80 (June 15): 411-415.

1828 Valls i Subirà, Oriol. 1963. "Arabian Paper in Catalonia: Notes on

Arabian Documents in the Royal Archives of the Kings of Aragon, in Barcelona." *The Paper Maker* 32 (March): 22-30.

1829 _____. 1965. "The Construction of the Mould in Catalonia During the Eighteenth Century." *The Paper Maker* 34 (September): 31-39.

1830 _____. 1972. "Estudio del Papel, Su Historia y Su Conservación." *Investigación y Técnica del Papel* 9 (October): 1117-1132.

1831 _____. 1971. "Estudio Sobre la Tuturación de los Trapos (1)." *Investigación y Técnica del Papel* 8 (April): 427-448.

1832 _____. 1970. "Estudio Sobre los Principios del Empleo de la Forma Con la Tela Fija, Seguido de Unas Notas Sobre los Primeros Intercambios Papeleros Entre Italia y Cataluña (1)." *Investigación y Técnica del Papel* 7 (April): 465-482.

1833 _____. 1968. "A Modo de Apéndice Al Trabajo del Sr. Basanta 'Filigranas en Documentos Gallegos'." *Investigación y Técnica del Papel* 5 (April): 421-425.

1834 _____. 1970. *Paper and Watermarks in Catalonia.* Amsterdam, The Netherlands: The Paper Publications Society. 2 Volumes.

1835 _____. 1965. "Three Hundred Years of Paper in Spain: From the Tenth to the Thirteenth Century." *The Paper Maker* 34 (March): 31-39.

1836 Vila, Pau. 1935. "L'Aspecte Geogràfic de la Indústria Paperera a Catalunya." *Butlletí del Centre Excursionista de Catalunya* 45 (No. 477): 68-76.

1837 _____. 1935. "L'Aspecte Geogràfic de la Indústria Paperera a Catalunya." *Butlletí del Centre Excursionista de Catalunya* 45 (No. 479): 144-151.

1838 _____. 1937. "Localisation et Évolution de l'Industrie du Papier en Catalogne." *Comptes Rendus du Congrès International de Géographie, Varsovie, 1934* 3: 301-311.

SWEDEN

1839 Althin, Torsten. 1959. "Ösjöfors, Schwedens Älteste Noch Bestehende Papiermühle." *Papiergeschichte* 9 (October): 43-48.

1840 Clemensson, Gustaf. 1923. *Klippans Papersbruk 1573-1923.* Lund, Sweden: Berlingska Boktryckeriet.

1841 Liljedahl, Gösta. 1956. "Eine Wasserzeichensammlung in Stockholm." *Papiergeschichte* 6 (November): 61-65.

1842   Platbarzdis, A. 1956. "Die Schwedischen Kartongeldscheine
        1716/17." *Papiergeschichte* 6 (July): 31-34.
1843   Voorn, Henk. 1959. "Early Papermaking in Sweden." *The Paper
        Maker* 28 (September): 1-8.
1844   _____. 1960. "The Story of Early Dutch Papermakers in
        Sweden." *The Paper Maker* 29 (February): 19-24.
1845   _____. 1960. "The Swedish Paper and Pulp Industry in the
        Nineteenth Century." *The Paper Maker* 29 (September):
        31-41.

SWITZERLAND

1846   Blaser, Fritz. 1955. "Die Älteste Papier-Urkunde des Staatsarchivs
        Luzern." *Papiergeschichte* 5 (November): 71.
1847   _____. 1974. "Brand der Papierfabrik Hartmann in Horw Bei
        Luzern." *Papiergeschichte* 23 (March): 17-19.
1848   _____. 1952. "Literatur Zur Geschichte der Schweiz.
        Papierfabrikation." In *The Briquet Album,* edited by E.J.
        Labarre. Hilversum, The Netherlands: The Paper Publications
        Society.. 74-78.
1849   _____. 1957. "Die Papiermühle Pfau in Schaffhausen."
        Papiergeschichte 7 (July): 56-58.
1850   _____. 1952. "Papiermühlen in den Vier Waldstätten."
        Papiergeschichte 2 (December): 82-84.
1851   _____. 1953. "Papiermühlen in den Vier Waldstätten."
        Papiergeschichte 3 (February): 8-11.
1852   _____. 1953. "Papiermühlen in den Vier Waldstätten."
        Papiergeschichte 3 (April): 27-28.
1853   _____. 1949. "Eine Projektierte Papiermühle." In *Buch und
        Papier,* edited by Horst Kunze. Leipzig, Germany: Otto
        Harrassowitz. 11-17.
1854   _____. 1971. "Die Schweizerischen Papiermühlen Ende 1798."
        Papiergeschichte 21 (December): 54-56.
1855   Fluri, Adolf. 1954. "Geschichte der Berner Papiermühlen."
        Papiergeschichte 4 (September): 47-52.
1856   _____. 1954. "Geschichte der Berner Papiermühlen."
        Papiergeschichte 4 (December): 79-84.
1857   _____. 1967. "Zur Geschichte der Berner Papiermühlen."
        Papiergeschichte 17 (June): 46-47.
1858   Frick, Gustav Adolf. 1923. "Die Schweizerische Papierfabrikation

Unter Besonderer Berücksichtigung des Sandortes." Unpublished Ph.D. dissertation, University of Bern.

1859   Häusler, Max. 1927. "Die Papiermühle und die Papierfabrik Auf dem Werd, 1472-1844." Unpublished Ph.D. dissertation, University of Zürich.

1860   Hössle, F. von. 1926. "Zur Entwicklung der Rheinischen Papierindustrie Einführung der Papiermaschine in Deutschland und der Schweiz." *Der Papier-Fabrikant* 24 (March 7): 148.

1861   _____. 1928. "Eine Papiermühlenwanderung von Biberach Nach Basel." *Wochenblatt für Papierfabrikation* 59 (June 9A): 10-27.

1862   Kälin, Hans. 1972. *Das Basler Papier-Gewerbe in der Reformationszeit.* Schinznach-Bad, Switzerland: Nejahrsgabe Schweizer Papier-Historiker.

1863   _____. 1973. *Vom Handel Mit Basler Papier im Mittelalter.* Schinznach-Bad, Switzerland: Nejahrsgabe Schweizer Papier-Historiker.

1864   Kazmeier, August Wilhelm. 1955. "Einige Daten Zur Tatigkeit Anton Galliziani's in Basel." *Gutenberg-Jahrbuch* 30: 16-18.

1865   Lanz, Werner. 1949. *Die Schweizerische Papierindustrie in Vergangenheit und Gegenwart.* Bern, Switzerland: Verband Schweizerischen Papier- und Papierstoff-Fabrikarten.

1866   Lindt, Johann. 1964. *The Paper-Mills of Berne and Their Watermarks 1465-1859.* Hilversum, The Netherlands: The Paper Publications Society.

1867   Naegeli, W. 1956. "Die Entstehungsgeschichte des Papiers." *Hespa Mitteilungen* 6: 11-19.

1868   Piccard, Gerhard. 1967. "Papiererzeugung und Buchdruck in Basel Bis Zum Beginn des 16. Jahrhunderts. Ein Wirtschaftsgeschichtlicher Beitrag." *Archiv für Geschichte des Buchwesens* 8: 25-322.

1869   Thiel, Viktor. 1938. "Papiererzeugung und Mountanindustrie in den Österr. Alpenländern." *Wochenblatt für Papierfabrikation* 69 (Special Edition): 9-15.

1870   Tschudin, Peter. 1956. "Das Basler Papiererhandwerk. Grundung und Entwicklung Bis 1530." *Stultifera Navis* 13 (October): 116-124.

1871   _____. 1958. "A History of Papermaking in Basel: Part I." *The Paper Maker* 27 (February): 33-39.

1872   _____. 1959. "A History of Papermaking in Basel: Part II." *The Paper Maker* 28 (February): 1-6.

1873 _____. 1962. "The Old Paper Mills of Switzerland and Their History: Part I." *The Paper Maker* 31 (September): 39-45.

1874 _____. 1963. "The Old Paper Mills of Switzerland and Their History: Part II." *The Paper Maker* 32 (March): 5-9.

1875 _____. 1963. "The Old Paper Mills of Switzerland and Their History: Part III." *The Paper Maker* 32 (September): 25-33.

1876 Tschudin, W.F. 1958. *The Ancient Paper-Mills of Basle and Their Marks.* Hilversum, The Netherlands: The Paper Publications Society.

1877 _____. 1950. "Etwas vom Papiermachen." *Schweizerisches Gutenbergmuseum* 36: 9-16.

1878 _____. 1959. "Ein Mainzer Fruhdruck Aus dem Jahre 1460 Auf Basler Papier." *Textil-Rundschau* 14 (December): 690-698.

1879 _____. 1960. "Von Zwei Alten Papiermacherwanderbüchern." *Textil-Rundschau* 15 (December): 637-646.

1880 Wyler, Edwin. 1927. "Die Geschichte des Basler Papiergewerbes." Unpublished Ph.D. dissertation, University of Basel.

WALES

1881 Davies, Alun E. 1967-1968. "Paper-Mills and Paper-Makers in Wales 1700-1900." *The National Library of Wales Journal* 15: 1-30.

1882 Kelleher, Denis. 1952. "Papermaking in Wales." *The Paper Maker* 21 (February): 51-56.

1883 Lewis, Peter W. 1968. "A Geography of the Papermaking Industry in England and Wales, 1860-1965." Unpublished Ph.D. dissertation, Manchester University.

1884 _____. 1969. *A Numerical Approach to the Location of Industry.* Hull, England: University of Hull Publications.

YUGOSLAVIA

1885 Kulundzic, Zvonimir. 1955. "Frühe Papiermacherei in Jugoslawien." *Papiergeschichte* 5 (July): 36-38.

1886 Ružić, Viktor. 1956. "Ein Blick in die Papiergeschichte Jugoslawiens." *Wochenblatt für Papierfabrikation* 84 (April 15): 232-233.

1887 _____. 1969. "Wortforschung in der Papiergeschichte Jugoslawiens." *Papiergeschichte* 19 (November): 28-30.

4 The History of Paper and Papermaking in North and South America

BRAZIL

1888   Bennett, R. 1938. "The Brazilian Paper Industry." *Paper Industry* 19 (January): 1166-1169.

1889   Radamus. 1936. "Die Entstehung der Brasilianischen Papier-industrie." *Wochenblatt für Papierfabrikation* 67 (January 18): 48.

1890   _____. 1936. "Die Entstehung der Brasilianischen Papier-industrie." *Wochenblatt für Papierfabrikation* 67 (March 21): 224-226.

1891   _____. 1936. "Die Entstehung der Brasilianischen Papier-industrie." *Wochenblatt für Papierfabrikation* 67 (May 2): 336-337.

1892   _____. 1936. "Die Entstehung der Brasilianischen Papier-industrie." *Wochenblatt für Papierfabrikation* 67 (October 3): 746-747.

1893   _____. 1937. "Die Entstehung der Brasilianischen Papier-industrie." *Wochenblatt für Papierfabrikation* 68 (June 5): 430-431.

1894   _____. 1937. "Die Entstehung der Brasilianischen Papier-industrie." *Wochenblatt für Papierfabrikation* 68 (September 25): 738-739.

1895   _____. 1938. "Die Entstehung der Brasilianischen Papier-industrie." *Wochenblatt für Papierfabrikation* 69 (May 7): 408-409.

1896   _____. 1938. "Die Entstehung der Brasilianischen Papier-industrie." *Wochenblatt für Papierfabrikation* 69 (July 30): 647-648.

1897   _____. 1938. "Die Entstehung der Brasilianischen Papier-industrie." *Wochenblatt für Papierfabrikation* 69 (August 27): 724-725.

1898   _____. 1938. "Die Entstehung der Brasilianischen Papier-industrie." *Wochenblatt für Papierfabrikation* 69 (September 3): 741-742.

1899   _____. 1939. "Die Entstehung der Brasilianischen Papier-industrie." *Wochenblatt für Papierfabrikation* 70 (August 12): 705-706.

CANADA

1900   Reilly, Desmond. 1952. "Early Papermaking in Canada." *The Paper Maker* 21 (February): 13-21.

MEXICO

1901   Asenjo Martínez, José Luis. 1968. "La Primera Fábrica de Papel Continuo en México." *Investigatión y Técnica del Papel* 5 (April): 289-296.

1902   Bockwitz, Hans H. 1949. "De Azteken en Maya-Indianen Als Papiermakers." *De Papierwereld* 3 (March): 358-360.

1903   _____. 1941. "Vom 'Papier' der Maya-Indianer." *Archiv für Buchgewerbe und Gebrauchsgraphik* 78 (July): 273-274.

1904   _____. 1939. "Vom 'Papier' der Maya-Indianer und den Maya-Hieroglyphen." *Wochenblatt für Papierfabrikation* 70 (July 29): 660-661.

1905   _____. 1949. "Die Papiermacherkunst und Ihre Bedeutung im Reiche der Azteken und Maya-Indianer." *Das Papier* 3 (February): 53-55.

1906   Christensen, Bodil. 1942. "Notas Sobre la Fabricacion del Papel Indigena y Su Empleo Para 'Brujerias' en la Sierra Norte de Puebla, Mexico." *Revista Mexicana de Estudios Antropologicos* 6 (January-August): 109-124.

1907   Godenne, Willy. 1960. "Le Papier des Comptes Communaux de Malines Datant du Moyen Âge." *Bulletin du Cercle Archéologique de Malines* 64: 36-53.

1908   Hunter, Dard. 1927. *Primitive Papermaking: an Account of a Mexican Sojourn and of a Voyage to the Pacific Islands in Search of Information, Implements and Specimens Relating to the Making & Decorating of Bark-Paper.* Chillicothe, Ohio: The Mountain House Press.

1909   Kalmár, Vera. 1957. "Mire Írtak Mexikó Őslakói?" *Papíripar és Magyar Grafika* 1 (July-August): 150-151.

1910   Lannik, William, Raymond L. Palm, and Marsha P. Tatkon. 1969. *Paper Figures and Folk Medicine Among the San Pablito Otomi.* New York: Museum of the American Indian, Heye Foundation.

1911   Lenz, Hans. 1959. "La Elaboración del Papel Indígena." *Esplendor del México Antiguo* 1: 355-360.

1912   _____. 1949. "Las Fibras y Plantas del Papel Indígena Mexicano." *Cuadernos Americanos* 8: 157-169.

1913   _____. 1959. "Loreto (Geschichte und Entwicklung Einer Papierfabrik)." *Papiergeschichte* 9 (April): 13-19.

1914   _____. 1959. "Loreto (Geschichte und Entwicklung Einer Papierfabrik)." *Papiergeschichte* 9 (August): 25-34.

1915 _____. 1957. *Loreto, Historia y Evolución de una Fabrica de Papel.* Mexico: Fábricas de Papel Loreto y Peña Pobre.

1916 _____. 1961. *Mexican Indian Paper; Its History and Survival.* Mexico: Editorial Libros de Mexico.

1917 _____. 1967. "Molinos Papeleros Mexicanos en la Época Colonial (1)." *Investigación y Técnica del Papel* 4 (January): 81-99.

1918 _____. 1950. *El Papel Indígena Mexicano, Historia y Supervivencia.* Mexico: Editorial Cultura.

1919 _____. 1968. "Das Papier in der Kolonialzeit Mexicos." *Papiergeschichte* 18 (July): 39-48.

1920 _____. 1969. "Papier und Aberglaube." *Papiergeschichte* 19 (December): 71-72.

1921 _____. 1970. "Papier und Aberglaube." *Papiergeschichte* 20 (June): 1-9.

1922 _____, and Federico Gómez de Orozco. 1940. *La Industria Papelera en México; Bosquejo Historico.* Mexico: Editorial Cultura.

1923 Miranda, Faustino. 1946. *Some Botanical Commentaries on Aztec Paper Making.* Mexico: Cuadernos Americanos.

1924 Reko, Victor A. 1934. "Altmexikanisches Papier." *Wochenblatt für Papierfabrikation* 65 (June 2): 386-388.

1925 _____. 1934. "Altmexikanisches Papier." *Wochenblatt für Papierfabrikation* 65 (June 9): 408-410.

1926 Richter, Oswald. 1938. "Untersuchungen an Papieren Aztekischer Völker Aus Kolumbischer und Vorkolumbischer Zeit und Über Chinesische, Türkische, Buddhistische Soghdische und Andere Papiere Aus den Turfan-Funden." *Faserforschung* 13: 57-81.

1927 Samayoa-Chinchilla, Carlos. 1969. "Le Papier Chez les Mayas." *Archeologia* No. 26 (January-February): 16-17.

1928 Sandermann, Wilhelm. 1970. "Papier in Altamerika." *Papiergeschichte* 20 (June): 11-24.

1929 _____. 1970. "Papier und Altamerika." *Papiergeschichte* 20 (October): 25-32.

1930 _____. 1969. "Papier und Bücher in Altamerikanischen Hochkulturen." *Vom Papier* No. 8: 20-33.

1931 Schwede, Rudolf. 1912. *Über das Papier der Maya-Codices und Einiger Altmexikanischer Bilderhandschriften.* Dresden, Germany: Verlag von Richard Bertling.

1932 _____. 1916. "Ein Weiterer Beitrag Zur Geschichte des
Altamerikanischen Papiers." *Jahresbericht der Vereinigung für
Angewandte Botanik* 13: 4-55.

1933 Starr, Frederick. 1900. "Mexican Paper." *American Antiquarian and
Oriental Journal* 22: 301-309.

1934 Tschudin, W.F. 1957. "O Staromexické Vyrobe Papiru." *Papír a
Celulósa* 12 (January): 21-23.

1935 Valentini, J.J. 1880. "Mexican Paper." *Proceedings of the American
Antiquarian Society* 1 (October): 58-81.

1936 Vindel, Francisco. 1956. *En Papel de Fabricación Azteca Fue
Impreso el Primer Libro en America.* Madrid, Spain: La
Academia Mexicana de la Historia.

1937 Von Hagen, Victor Wolfgang. 1943. *The Aztec and Maya Papermakers.* New York: J.J. Augustin Publisher.

1938 _____. 1943. "Mexican Papermaking Plants." *Journal of the
New York Botanical Garden* 45 (January): 1-10.

UNITED STATES

1939 Adams, W. Claude. 1951. "History of Papermaking in the Pacific
Northwest: I." *Oregon Historical Quarterly* 52 (March): 21-37.

1940 _____. 1951. "History of Papermaking in the Pacific Northwest:
II." *Oregon Historical Quarterly* 52 (June): 83-100.

1941 _____. 1951. "History of Papermaking in the Pacific Northwest:
III." *Oregon Historical Quarterly* 52 (September): 154-185.

1942 _____. 1951. *History of Papermaking in the Pacific Northwest.*
Portland, Oregon: Binfords & Mort.

1943 Allen, George. 1942. "The Rittenhouse Paper Mill and Its Founder."
Mennonite Quarterly Review 16 (April): 108-128.

1944 Allen, J.H. 1937. "History of Pulp and Paper in the South." *Paper
Trade Journal* 65 (October 28): 133-134.

1945 Babler, Otto F. 1934. "Benjamin Franklin und das Papier."
Zeitschrift für Büchfreunde (Third Series) 3 (March): 59.

1946 Barker, Charles R. 1926. "Old Mills of Mill Creek, Lower Merion."
The Pennsylvania Magazine of History and Biography 50
(No. 1): 1-22.

1947 Beach, H.A. 1932. "The Ancient Craft of Hand-Made Paper." *The
Publisher's Weekly* 122 (October 1): 1343-1350.

1948 Beatty, William B. 1959. "Early Papermaking in Utah." *The Paper
Maker* 28 (February): 9-20.

1949   Buehler, J. Marshall. 1966. "Birthplace of the Wisconsin Paper
        Industry." *The Paper Maker* 35 (October): 13-25.
1950   Chapin, Howard M. 1926. "Early Rhode Island Paper Making." *The
        Americana Collector* 2 (May): 303-309.
1951   Coulter, E. Merton. 1964. "A Note on a Georgia Paper Mill." *The
        Georgia Historical Quarterly* 48 (June): 239-242.
1952   Crane, Ellery B. 1887. *Early Paper Mills in Massachusetts, Especially
        Worcester County.* Worcester, Massachusetts: Worcester Print
        & Publishing Company.
1953   Cunz, Dieter. 1945. "Maryland's First Papermaker." *The American-
        German Review* 12: 21-23.
1954   Denis, Leonard. 1961. "Hanwell Paper: a Nineteenth-Century
        Wissahickon Valley Industry." *The Paper Maker* 30
        (September): 45-49.
1955   Dickoré, Marie Paula. 1947. "The Wadsworth Paper Mill; a Water-
        mark Furnishes a Clue to the History of the First Paper Mill in
        the Little Miami Valley." *Bulletin of the Historical and
        Philosophical Society of Ohio* 5 (March): 7-24.
1956   Dill, Alonzo Thomas. 1968. *Chesapeake, Pioneer Papermaker: a
        History of the Company and Its Community.* Charlottesville,
        Virginia: The University Press of Virginia.
1957   Donnelly, Florence. 1951. "The Beautiful Mill." *The Paper Maker*
        20 (February): 23-32.
1958   _____ . 1960. "Comas Paper Mill: First in Washington." *The
        Paper Maker* 29 (September): 12-28.
1959   _____ . 1950. "First in the West." *The Paper Maker* 19
        (September): 1-9.
1960   _____ . 1958. "Oregon's Pioneer Paper Mill: First in the North-
        west." *The Paper Maker* 27 (February): 13-22.
1961   _____ . 1958. "Oregon's Second Venture in Papermaking: the
        Clackamas Mill." *The Paper Maker* 27 (September): 21-28.
1962   _____ . 1949. "Pioneer Paper Mill of the West." *The Paper Maker*
        18 (September): 15-19.
1963   _____ . 1951. "The Pride of the San Joaquin." *The Paper Maker*
        20 (September): 33-41.
1964   _____ . 1953. "The San Lorenzo Paper Mill: Where the Red-
        woods Met the Sea." *The Paper Maker* 22 (February): 11-23.
1965   _____ . 1954. "Saratoga Paper Mill." *The Paper Maker* 23
        (February): 29-39.
1966   _____ . 1956. "Trail of Ventures." *The Paper Maker* 25
        (February): 17-27.

1967   Dugan, Frances L.S. 1959. "A History of Papermaking in
        Kentucky." *Papiergeschichte* 9 (November): 54-62.
1968   ————. 1960. "A History of Papermaking in Kentucky."
        Papiergeschichte 10 (February): 4-12.
1969   ————, and Jacqueline P. Bull, editors. 1959. *Bluegrass Crafts-
        man: Being the Reminiscences of Ebenezer Hiram Stedman,
        Papermaker 1808-1885.* Lexington, Kentucky: University of
        Kentucky Press.
1970   Edelstein, Sidney M. 1961. "Papermaker Joshua Gilpin Introduces
        the Chemical Approach to Papermaking in the United States."
        The Paper Maker 30 (September): 3-12.
1971   Edwards, Frances. 1966. "Connecticut Paper Mills: the Eagle Mill in
        Suffield." *The Paper Maker* 35 (October): 5-9.
1972   ————. 1966. "Connecticut Paper Mills: the Franklin Mill in
        Suffield." *The Paper Maker* 35 (June): 11-16.
1973   ————. 1967. "Connecticut Paper Mills: the Leffingwell Mill at
        Norwich, First in the Connecticut Colony." *The Paper Maker*
        36 (September): 22-25.
1974   Elliott, Harrison. 1951. "Benjamin Franklin: Paper Mill Promoter
        and Patron." *The Paper Maker* 20 (February): 35-40.
1975   ————. 1952. "A Century Ago an Eminent Author Looked Upon
        Paper and Papermaking." *The Paper Maker* 21 (September):
        55-58.
1976   ————. 1950. "Connecticut's First Papermaker." *The Paper
        Maker* 19 (September): 41-43.
1977   ————. 1950. "Cyrus W. Field: a Successful Paper Merchant."
        The Paper Maker 19 (February): 7-9.
1978   ————. 1953. "The First Paper Mill in New York." *The Paper
        Maker* 22 (September): 25-30.
1979   ————. 1950. "The Gilpin Paper Mill on the Brandywine." *The
        Paper Maker* 19 (February): 25-26.
1980   ————. 1949. "Some Paper Mill Labels of Antiquarian Interest
        Designed and Engraved By Alexander Anderson, the First
        Wood Engraver in This Country." *The Paper Maker* 18
        (February): 7-12.
1981   Gilpin, Thomas, Jr. 1925. "Memoir of Thomas Gilpin." *The
        Pennsylvania Magazine of History and Biography* 49 (No. 4):
        289-328.
1982   Goerl, Stephen. 1945. *Papermaking in America.* New York: Bulkley.
1983   Goodwin, Rutherfoord. 1937. "The Williamsburg Paper Mill of

William Parks, the Printer." *The Papers of the Bibliographical Society of America* 31 (No. 2): 21-44.

1984   Goold, William. 1874. *Early Papermills of New-England.* Bath, Maine: Maine Historical Society.

1985   Halley, R.A. 1904. "Paper Making in Tennessee." *The American Historical Magazine* 9 (July): 211-217.

1986   Hancock, Harold B. 1959. "An American Papermaker in Europe—1795-1801." *The Paper Maker* 28 (September): 11-15.

1987   _____, and Norman B. Wilkinson. 1957. "The Gilpins and Their Endless Papermaking Machine." *The Pennsylvania Magazine of History and Biography* 81 (October): 391-405.

1988   _____, and _____. 1958. "Thomas and Joshua Gilpin, Papermakers." *The Paper Maker* 27 (September): 1-11.

1989   Hayden, Seth. 1968. "Papermaking in Chambersburg, Pennsylvania." *The Paper Maker* 37 (September): 3-9.

1990   Hommel, Rudolf P. 1947. "Two Centuries of Papermaking at Miquon, Pennsylvania." *Bulletin of the Historical Society of Montgomery County (Pennsylvania)* 5 (April): 275-290.

1991   Hosler, Wilbert. 1938. "History of Papermaking in Michigan." *Michigan Historical Magazine* 22 (Autumn): 361-402.

1992   Hössle, F. von. 1930. "Eine Papiermühlen- und Koloniegründung Durch Deutsche Papiermacher in Nordamerika." *Wochenblatt für Papierfabrikation* 61 (April 5): 443-445.

1993   Hunter, Dard. 1951. "American Paper Labels." *Gutenberg-Jahrbuch* 26: 30-33.

1994   _____. 1952. "The Early Paper Mills of Ohio." In *The Briquet Album,* edited by E.J. Labarre. Hilversum, The Netherlands: The Paper Publications Society. 85-96.

1995   _____. 1940. "Lost and Not Lost: Handmade Paper, Rich in Historical Associations, Has Special Meaning on Its 250th Anniversary in America." *Technology Review* 42 (January): 109-111, 124, 126.

1996   _____. 1946. "Ohio's Pioneer Paper Mills." *Paper Industry* 28 (April): 98, 100, 102, 104.

1997   _____. 1950. "Papermaking By Hand in America." *Gutenberg-Jahrbuch* 25: 31-40.

1998   _____. 1950. *Papermaking By Hand in America.* Chillicothe, Ohio: The Mountain House Press.

1999   _____. 1952. *Papermaking in Pioneer America.* Philadelphia, Pennsylvania: University of Pennsylvania Press.

2000   James, Arthur E. 1970. "The Paper Mills of Chester County Penn-
        sylvania 1779-1967: Part I." *The Paper Maker* 39 (September):
        3-18.

2001   Jones, Horatio Gates. 1896. "Historical Sketch of the Rittenhouse
        Paper-Mill; the First Erected in America, A.D. 1690." *The
        Pennsylvania Magazine of History and Biography* 20 (No. 3):
        315-333.

2002   Leonard, Eugenie Andruss. 1950. "Paper As a Critical Commodity
        During the American Revolution." *The Pennsylvania Magazine
        of History and Biography* 74 (October): 488-499.

2003   Lockwood Trade Journal Co., Inc. 1940. *250 Years of Papermaking
        in America.* New York: Lockwood Trade Journal Co., Inc.

2004   McCorison, Marcus A. 1963. "Vermont Papermaking 1784-1820."
        Vermont History 31 (October): 209-245.

2005   _____. 1964. "Vermont Papermaking 1784-1820: Part One."
        The Paper Maker 33 (March): 19-28.

2006   _____. 1964. "Vermont Papermaking 1784-1820: Part Two."
        The Paper Maker 33 (September): 23-31.

2007   McRae, J. Finley. 1956. *Paper Making in Alabama.* Downingtown,
        Pennsylvania: The Newcomen Society.

2008   Magee, J.F. 1934. "Watermarks of Early American Paper Makers."
        Paper Trade Journal 98 (May 24): 43-44.

2009   Mason, John. 1962. The Bird & Bull Press of Henry Morris." *The
        Black Art* 1 (Winter): 114-118.

2010   Maxson, John W., Jr. 1962. "American Papermakers in the Great
        Tariff Debate." *The Paper Maker* 31 (March): 14-34.

2011   _____. 1961. "Coleman Sellers: Machine Maker to America's
        First Mechanized Paper Mills." *The Paper Maker* 30
        (February): 13-27.

2012   _____. 1969. "George Escol Sellers: Inventor, Historian, and
        Papermaker." *The Paper Maker* 38 (June): 40-57.

2013   _____. 1960. "Nathan Sellers: America's First Large-Scale Maker
        of Paper Moulds." *The Paper Maker* 29 (February): 1-15.

2014   _____. 1968. "Papermaking in America: From Art to Industry,
        1690 to 1860." *The Quarterly Journal of the Library of
        Congress* 25 (April): 116-129.

2015   Munnikhuysen, John B. 1939. *A Short History of the Paper Making
        Industry in Baltimore City and County, Maryland.* Baltimore,
        Maryland: Enoch Pratt Library.

2016   Newton, Janet Foster. 1942. "Early American Papermakers: From
        Wasp to Watermark." *Antiques* 42 (September): 132-133.

2017   Nugent, John M. 1957. "The Paper Mills of Lower Merion." *The Paper Maker* 26 (February): 19-27.

2018   Pennypacker, S.W. 1899. "William Rittenhouse and the Paper Mill." In *The Settlement of Germantown, Pennsylvania, and the Beginning of German Emigration to North-America,* by S.W. Pennypacker. Philadelphia, Pennsylvania: J. Campbell. 162-168.

2019   Renker, Armin. 1939. "Benjamin Franklin und Chinesische Papier." *Der Papier-Fabrikant* 37 (January 6): 15.

2020   _____. 1932. "Bericht Über Neue Werkdruckpapiere für Amerika." *Imprimatur* 3: 192-193.

2021   _____. 1959. "William Cobbett and His Cornstalk Paper." *The Paper Maker* 28 (February): 23-30.

2022   _____. 1939. "Zweihundertfünfzig Jahre Papiermacherei in den Vereinigten Statten von Amerika. Wilhelm Rittenghausen Aus Mülheim (Ruhr) Ihr Begründer." *Gutenberg-Jahrbuch* 14: 39-46.

2023   Rockefeller, George C. 1953. "Early Paper-Making in Trenton: Some Accounts of the Potts & Reynolds Mill 1777-1785." *Proceedings of the New Jersey Historical Society* 71 (January): 24-32.

2024   Rubincam, Milton. 1941. Bericht Über die Feier des 250. Jahrestages der USA.-Papiermacherei." *Wochenblatt für Papierfabrikation* 72 (May 31): 327-328.

2025   Schrohe, A. 1927. "Zwei Deutsche Pioniere der Papierindustrie in Nordamerika." *Der Papier-Fabrikant* 25 (March 13): 166-167.

2026   Schulte, A. 1940. Heimat und Name von Wilhelm Rittinghausen, dem Ersten Papiermacher der Vereinigten Staaten." *Wochenblatt für Papierfabrikation* 71 (November 30): 635-639.

2027   Seitz, May A. 1946. *The History of the Hoffman Paper Mills in Maryland.* Baltimore, Maryland: Holliday Press.

2028   Slaff, W. 1944. "Romance of the Paper Watermark." *American Paper Merchant* 41 (June): 26.

2029   Smith, Albert Barton. 1938. *From Spook Hill to Loveland in 1810. Historical Paper Mill Sketches By an Old American Papermaker.* Trenton, New Jersey: Privately Printed.

2030   Smith, David C. 1970. *History of Papermaking in the United States (1691-1969).* New York: Lockwood Publishing Company.

2031   Smith, J.E.A. n.d. *Pioneer Papermaking in Berkshire. The Life, Life*

Work and Influence of Zenas Crane. Holyoke, Massachusetts: Privately Printed.

2032 Snell, Ralph M. 1932. "An Early American Paper Mill." *The Paper Maker* 1 (June-July): 15-16.

2033 _____. 1933. "Paper Making at Chambersburg, Pennsylvania." *The Paper Maker* 2 (November-December): 113-115, 119.

2034 _____. 1934. "Paper Making at Chambersburg, Pennsylvania." *The Paper Maker* 3 (January-March): 11-13, 19.

2035 _____. 1932. "Paper Making at Niagara Falls." *The Paper Maker* 1 (October-November): 40-41.

2036 _____. 1933. "Paper Making at Niagara Falls." *The Paper Maker* 2 (January-February): 16-18.

2037 _____. 1933. "Paper Making at Niagara Falls." *The Paper Maker* 2 (May-June): 51-54.

2038 _____. 1933. "Paper Making on the Skaneateles Outlet." *The Paper Maker* 2 (July-August): 68-71.

2039 _____. 1933. "Paper Making on the Skaneateles Outlet." *The Paper Maker* 2 (September-October): 93-96.

2040 _____. 1932. "Pioneer Pulp and Paper Making at Niagara Falls." *The Paper Maker* 1 (August-September): 24-26.

2041 _____. 1934. "The Sellers Family and the Story of Paper Making at Sellers Landing, Illinois." *The Paper Maker* 3 (April-June): 30-32, 43.

2042 _____. 1934. "The Sellers Family and the Story of Paper Making at Sellers Landing, Illinois." *The Paper Maker* 3 (July-September): 57-60, 68.

2043 _____, and B.T. McBain. 1934. "Early Pulp and Paper Mills of the Pacific Coast." *Paper Trade Journal* 99 (October 11): 42-50.

2044 Sullivan, Frank. 1950. "The Most Expensive Paper in America." *The Paper Maker* 19 (September): 33-39.

2045 _____. 1951. "Paper, Papermaking, and the Patriots: Notes to Accompany the Papermakers March to the Revolutionary War." *The Paper Maker* 20 (September): 27-31.

2046 Thayer, C.S. 1970. "Patriot Papermakers and Their Watermarks." *Modern Lithography* 38 (August): 26-27, 30-31.

2047 Watson, B.G. 1966. "Bryan Donkin: Pioneer Paper-Machine Manufacturer." *The Paper Maker* 35 (June): 21-32.

2048 Weeks, Lyman Horace. 1916. *A History of Paper-Manufacturing in the United States, 1690-1916.* New York: The Lockwood Trade Journal Company.

2049   Weston, Harry E. 1944. "A Chronology of Papermaking in the United States: Part I." *The Paper Maker* 13 (September): 9-11.

2050   _____. 1945. "A Chronology of Papermaking in the United States." *The Paper Maker* 14 (February): 8-10.

2051   Wheelwright, William Bond. 1952. "A Chronicle of Bird & Son, Inc." *The Paper Maker* 21 (September): 37-46.

2052   _____. 1941. "The Gilpins of Delaware." *The Paper Maker* 10 (February): 6-11.

2053   _____. 1940. "Pioneers of American Paper Making." *The Paper Maker* 9 (February): 16-20.

2054   _____. 1941. "A Strange Chapter in Southern Papermaking." *The Paper Maker* 10 (September): 2-5.

2055   _____. 1942. "War and the Paper Industry in 1776." *The Paper Maker* 11 (February): 3-6.

2056   _____. 1951. "Zenas Crane, Pioneer Papermaker." *The Paper Maker* 20 (February): 1-7.

2057   Willcox, Joseph. 1911. *Ivy Mills 1729-1866*. Baltimore, Maryland: Privately Printed by Lucas Brothers, Inc.

2058   Wiswall, Calrence Augustus. 1938. *One Hundred Years of Paper Making; a History of the Industry on the Charles River at Newton Lower Falls, Massachusetts*. Reading, Massachusetts: Reading Chronicle Press, Inc.

5   *The Study*
     of Paper History

PAPER HISTORY RESEARCH

2059 Bockwitz, Hans H. 1940. "Die 'Forschungsstelle Papiergeschichte' im Ersten Jahre Ihres Bestehens." *Archiv für Buchgewerbe und Gebrauchsgraphik* 77 (May): 159-161.

2060 _____. 1949. "Fortschritte in der Papiergeschichtsforschung." *Wochenblatt für Papierfabrikation* 77 (September): 320.

2061 _____. 1938. "Neue Beiträge Zur Papiergeschichte im Gutenberg-Jahrbuch für 1938." *Wochenblatt für Papierfabrikation* 69 (August 6): 668-669.

2062 _____. 1941. "Neue Ergebnisse der Papiergeschichtsforschung." *Archiv für Buchgewerbe und Gebruachsgraphik* 78 (May): 182-183.

2063 _____. 1940. "Neue Ergebnisse der Papiergeschichtsforschung im Gutenberg-Jahrbuch 1940." *Wochenblatt für Papierfabrikation* 71 (December 21): 701-702.

2064 _____. 1942. "Neue Ergebnisse der Papiergeschichtsforschung im Gutenberg-Jahrbuch 1941." *Wochenblatt für Papierfabrikation* 73 (April 18): 118.

2065 _____. 1939. "Neue Forschungen Zur Geschichte des Papiers im Gutenberg-Jahrbuch 1939." *Wochenblatt für Papierfabrikation* 70 (August 26): 745-747.

2066 _____. 1938. "Neuere Papiergeschichtliche Literatur." *Archiv für Buchgewerbe und Gebrauchsgraphik* 75 (September): 363-366.

2067 _____. 1938. "Papiergeschichte Als Wissenschaft." *Archiv für Buchgewerbe und Gebrauchsgraphik* 75 (October): 407.

2068 _____. 1941. "Die Papiergeschichtliche Literatur des Jahres 1939." *Wochenblatt für Papierfabrikation* 72 (May 10): 287.

2069 _____. 1950. "Papiergeschichtsforschung in Holland." *Wochenblatt für Papierfabrikation* 78 (June 15): 309.

2070 _____. 1937. "Zukunftsaufgaben der Papiergeschichtsforschung." *Wochenblatt für Papierfabrikation* 68 (October 23): 815-816.

2071 Bogdán, István. 1956. "Vizjelek és Vizjelkutatás." *Livéltári Hiradó* 8: 27-35.

2072 _____. 1956. "Vizjelek és Vizjelkutatás." *Papír- és Nyomdatechnika* 8 (January): 25-28.

2073 _____. 1959. "A Vizjelkutatás Problémái. Vizjelgyüjtésünk Módszertaha." *Levéltári Közlemények* 30: 89-108.

2074 Fiskaa, H.M. 1968. "Filigranologi. En Hjelpevitenskap i Vekst." *Bibliotek og Forskning* 17: 52-67.

139

2075 ———. 1965. "Filigranologie. Eine Heranwachsende Hilfwissenschaft." *Papiergeschichte* 19 (November): 39-45.

2076 Gasparinetti, A.F. 1952. "Per l'Adozione Di una Terminologia Generale Delle Filigrane." In *The Birquet Album,* edited by E.J. Labarre. Hilversum, The Netherlands: The Paper Publications Society. 119-121.

2077 ———. 1953. "Über ein Systematisches Studium der Wasserzeichen." *Papiergeschichte* 3 (September): 49-51.

2078 ———. 1953. "Über eine Allgemein Gültige Terminologie für die Wasserzeichen." *Papiergeschichte* 3 (September): 46-47.

2079 Gerardy, Theodor. 1964. *Datieren Mit Hilfe von Wasserzeichen; Beispielhaft Dargestelt an der Gesamtproduktion der Schaumbrugischen Papiermühle Ahrensburg von 1604-1650.* Bückeburg, Germany: Verlag Grimme Bückeburg.

2080 ———. 1969. "Der Identitätsbeweis Bei der Wasserzeichendatierung." *Archiv für Geschichte des Buchwesens* 9: 733-778.

2081 ———. 1956. "Zur Methodik der Wasserzeichenforschung." *Papiergeschichte* 6 (April): 14-20.

2082 ———. 1959. "Probleme der Wasserzeichenforschung." *Papiergeschichte* 9 (December): 66-73.

2083 ———. 1962. "Zur Terminologie der Wasserzeichenkunde (Korreferat)." *Papiergeschichte* 12 (February): 17-18.

2084 Irigoin, Jean. 1971. "Quelques Méthodes Scientifiques Applicables à l'Etude Historique du Papier." *Papiergeschichte* 21 (October): 4-9.

2085 Kazmeier, August Wilhelm. 1949. "Gragen Zur Frühesten Papiermacherei im Alten Mittelamerika und Einige Probleme Unserer Heutigen Wasserzeichenforschung." *Das Papier* 3 (December): 458-461.

2086 ———. 1950. "Die Regionale Entwicklung der Wasserzeichenforschung im Umriss Dargestellt." *Gutenberg-Jahrbuch* 25: 25-30.

2087 Langenbach, Alma. 1951. "Das Wasserzeichen Als Objekt der Volkskundlichen Forschung." *Gutenberg-Jahrbuch* 26: 22-29.

2088 Liljedahl, Gösta. 1961. "Filigranology at Present in Sweden." *Papiergeschichte* 11 (December): 87-89

2089 ———. 1968. "Om Pappershistoria Och Filigranologi 1958-1968." *Historisk Tidskrift (Second Series)* 31 (October-December): 480-502.

2090 _____. 1970. "Om Vattenmärken i Paper Och Vattenmärksforskning (Filigranologi)." *Biblis* n.v.: 91-129.

2091 Maleczyńska, Kazimiera. 1967. "Badania Nad Dawnym Papiernictwem w Polsce w Latach 1944-1965." *Roczniki Biblioteczne* 11 (Nos. 1-2): 193-200.

2092 Mošin, Vladimir. 1955. "Die Evidentierung und Datierung der Wasserzeichen." *Papiergeschichte* 5 (September): 49-57.

2093 _____. 1974. "Die Filigranologie Als Historische Hilfswissenschaft." *Papiergeschichte* 23 (June): 29-48.

2094 _____. 1954. "Filigranologija Kao Pomočna Historijska Nauka." *Zbornik Historijskog Instituta Jugoslavenske Akademije* 1: 70-83.

2095 Renker, Armin. 1938. "Die Forschungsstelle Papiergeschichte in Mainz." *Wochenblatt für Papierfabrikation* 69 (December): 1104-1107.

2096 Schlieder, Wolfgang. 1973. "Výnam Historických Výzkumů Pro Papírenskou Současnost." *Papír a Celulósa* 28 (December): 267.

2097 Schulte, Toni. 1961. "Neue Erkenntnisse Zur Papiergeschichtlichen Ikonographie." *Papiergeschichte* 11 (February): 13-20.

2098 Stevenson, Allan. 1961. *Observations on Paper As Evidence.* Lawrence, Kansas: University of Kansas Publications Library Series No. 11.

2099 _____. 1962. "Paper As Bibliographical Evidence." *The Library (Fifth Series)* 17: 197-212.

2100 Tanselle, G. Thomas. 1971. "The Bibliographical Description of Paper." *Studies in Bibliography* 24: 27-67.

2101 Tschudin, W.F. 1964. "Stand der Forschung Über die Schweiz. Papiermühlen, Papier- und Kartonfabriken und Deren Marken Zur Zeit der Schweizerischen Landesausstellung 1964." *Textil-Rundschau* 19 (March): 147-157.

2102 _____. 1964. "Stand der Forschung Über die Schweiz. Papiermühlen, Papier- und Kartonfabriken und Deren Marken Zur Zeit der Schweizerischen Landesausstellung 1964 (II. Teil)." *Textil-Rundschau* 19 (April): 208-216.

2103 _____. 1964. "Stand der Forschung Über die Schweiz. Papiermühlen, Papier- und Kartonfabriken und Deren Marken Zur Zeit der Schweizerischen Landesausstellung 1964 (III. Teil)." *Textil-Rundschau* 19 (May): 263-273.

2104 Weiss, Karl Theodor. 1950. "Die Bedeutung des Gesetzes der

Formenpaare für die Wasserzeichenkunde." *Allgemeine Papier-Rundschau* No. 4 (February 28): 164-166.

2105 Weiss, Wisso. 1961. "Bemerkungen Über Begriff, Forschungs-methode und Aufgaben der Wasserzeichenkunde." *Papiergeschichte* 11 (February): 5-8.

2106 ————. 1962. "Zur Terminologie der Wasserzeichenkunde." *Papiergeschichte* 12 (February): 9-17.

2107 White, F.A. 1914. "A Contribution to the Study of Watermarks." *Paper Makers' Monthly Journal* 52 (September 15): 306-308.

PAPER HISTORIANS

2108 Alibaux, Henri. 1952. "Méthode de Travail de Charles Moïse Briquet." In *The Briquet Album,* edited by E.J. Labarre. Hilversum, The Netherlands: The Paper Publications Society.

2109 Benzing, Josef. 1955. "Hans Lens und Seine Forschungen Zur Mexikanischen Papiergeschichte." *Papiergeschichte* 5 (July): 29-35.

2110 Blaser, Fritz. 1952. "Charles Moïse Briquet und die Schweizerische Papiergeschichtsforschung. Eine Bibliographische Studie." In *The Briquet Album,* edited by E.J. Labarre. Hilversum, The Netherlands: The Paper Publications Society. 69-73.

2111 ————. 1959. "Samuel Engel und die Wasserzeichenforschung." *Papiergeschichte* 9 (December): 78-80.

2112 ————. 1963. "Schweizer Papierforscher Neben Charles Moïse Briquet." *Papiergeschichte* 13 (April): 7-11.

2113 Bockwitz, Hans H. 1937. "Armin Renkers Literarisches Wirken im Dienste der Papierforschung." *Sankt Wiborada 4: 157.*

2114 ————. 1948. "E.J. Labarre, ein Englischer Papierforscher in Holland." *Wochenblatt für Papierfabrikation* 76 (July): 171.

2115 ————. 1952. "Zum 50. Todestage von Friedrich Keinz." In *The Briquet Album,* edited by E.J. Labarre. Hilversum, The Netherlands: The Paper Publications Society. 79-82.

2116 ————. 1938. "Zur Französischen Papiergeschichtsforschung der Gegenwart. Henri Alibaux, Lyon und Seine Papiergeschicht-lichen Arbeiten." *Wochenblatt für Papierfabrikation* 69 (July 23): 628-629.

2117 ————. 1949. "Henri Alibaux Als Papierforscher, 1872-1941." *Wochenblatt für Papierfabrikation* 77 (January): 17-18.

2118 ————. 1938. "Zu Karabaceks Forschungen Über das Papier im Islamischen Kulturkreis." *Buch und Schrift* 1: 83-86.

2119   Crawford, Nelson Antrim. 1924. "The Books of Dard Hunter." *The American Mercury* 2 (August): 470-472.

2120   Elliott, Harrison. 1926. "A Visit to Dard Hunter." *The American Printer* 82 (March 20): 27-30.

2121   Gachet, Henri. 1957. "Monsieur des Billettes, Historien du Papier." *Papiergeschichte* 7 (July): 49-50.

2122   Gasparinetti, A.F. 1964. "Another Footnote to Paper History." *The Paper Maker* 33 (March): 45-51.

2123   _____. 1959. "F.M. Nigrisoli Als Papiergeschichtsforscher." *Papiergeschichte* 9 (December): 80-83.

2124   _____. 1955. "A Historical Contribution to C.M. Briquet." *The Paper Maker* 24 (September): 49-51.

2125   _____. 1962. "Some Footnotes to the Work of C.M. Briquet." *The Paper Maker* 31 (March): 61-63.

2126   Günter, Herbert. 1961-1962. "Armin Renker Zum Gedächtnis." *Imprimatur (New Series)* 3: 262-266.

2127   Hunter, Dard. 1941. *Before Life Began, 1883-1923.* Cleveland, Ohio: The Rowfant Club.

2128   _____. 1936. "Making Books on Paper Making." *Paper Mill and Wood Pulp News* 59 (April): 15-16, 20-21.

2129   _____. 1958. *My Life With Paper: an Autobiography.* New York: Alfred Knopf.

2130   Johnson, Fridolf. 1967. "Henry Morris: Papermaker and Printer." *American Artist* 31 (October): 56-61, 84-87.

2131   Philpott, A.J. 1934. *A Lesson in Papermaking By Dard Hunter.* Boston, Massachusetts: The Club of Odd Volumes.

2132   Renker, Armin. 1933. "Dard Hunter: Erinnerungen und Ausblicke." *Imprimatur* 4: 56-64.

2133   _____. 1932. "Dr. Karl Theodor Weiss in Mönchweiler und Seine Papiergeschichtliche Sammlung." *Zeitschrift für Büchfreunde (Third Series)* 1 (July): 147-152.

2134   _____. 1944-1949. "Drei Forscher: Zum Andenken an die Papiergeschichtsforscher Henri Alibaux, Alfred Schulte, Dr. Karl Theodor Weiss. Mit Drei Abbildungen Auf Einer Kunstdrucktafil." *Gutenberg-Jahrbuch* 19-24: 47-52.

2135   _____. 1931. "Leben und Schicksal des Wasserzeichenforschers Charles-Moïse Briquet." *Philobiblon* 4 (January): 19-22.

2136   _____. 1931. "Leben und Schicksal des Wasserzeichenforschers Charles-Moïse Briquet." *Philobiblon* 4 (February): 67-69.

2137   _____. 1931. "Leben und Schicksal des Wasserzeichenforschers Charles-Moïse Briquet." *Philobiblon* 4 (March): 103-104.

2138 _____ . 1952. "Leben und Schicksal des Wasserzeichenforschers Charles-Moïse Briquet." In *The Briquet Album,* edited by E.J. Labarre. Hilversum, The Netherlands: The Paper Publications Society. 13-25.

2139 _____ . 1936. "Der Papierforscher." *Wochenblatt für Papierfabrikation* 67 (December 5): 915-918.

2140 Schulte, Alfred. 1952. "C.M. Briquet's Werk und die Aufgaben Seiner Nachfolger." In *The Briquet Album,* edited by E.J. Labarre. Hilversum, The Netherlands: The Paper Publications Society. 49-54.

2141 Siberell, Lloyd Emerson. 1935. "Dard Hunter, the Mountain House and Chillicothe." *Ohio State Archaeological and Historical Quarterly* 44 (April): 238-244.

2142 _____ . 1935. *Dard Hunter, the Mountain House and Chillicothe.* Cincinnati, Ohio: Ailanthus Press.

2143 Voorn, Henk. 1968. "Een Amerikaanse Drukker Maakt Zijn Eigen Papier." *De Papierwereld* 23 (January): 7-8.

2144 _____ . 1950. "Dard Hunter." *De Papierwereld* 4 (March): 306-308.

2145 _____ . 1952. "Germany's Contribution to the Study of Paper-making History." *The Paper Maker* 21 (September): 50-52.

2146 Zuman, F. 1934. "Aus der Geschichte der Papiermacherei." *Böhmische Historische Zeitschrift* 20: 675-678.

PAPER MUSEUMS

2147 Bockwitz, Hans H. 1939. "Ein Papier-Museum in Cambridge (Massachusetts)." *Wochenblatt für Papierfabrikation* 70 (July 29): 659-660.

2148 Hunter, Dard. 196-. *The Dard Hunter Paper Museum.* Appleton, Wisconsin: Institue For Paper Chemistry.

2149 _____ . 1941. "Dard Hunter Paper Museum of the Massachusetts Institute of Technology." *Print* 2 (October-December): 43-48.

2150 Klemm, Johannes. 1932. "Die Abteilung Papier im Deutschen Museum zu Munchen." *Wochenblatt für Papierfabrikation* 63 (June 4A): 5-8.

2151 Libiszowski, Stefan. 1971. "Papiernia - Muzeum." *Przegląd Papierniczy* 27 (February): 53-56.

2152 Marchlewska, Jadwiga. 1973. "Niemieckie Muzeum Papieru." *Przegląd Papierniczy* 29 (January): 22-28.

2153   ———. 1972. "Das Polnische Papiermuseum in Duszniki (Reinerz)." *Papiergeschichte* 22 (June): 27-29.

2154   Narita, Kiyofusa. 1954. "Paper Museum in Tokyo (Japan)." *Papiergeschichte* 4 (July): 29-32.

2155   Renker, Armin. 1938. "Eine Forschungsstelle für Papiergeschichte im Gutenberg-Museum in Mainz." *Imprimatur* 8: 172-173.

2156   ———. 1949. "Die Forschungsstelle Papiergeschichte im Gutenberg-Museum zu Mainz." In *Buch und Papier,* edited by Horst Kunze. Leipzig, Germany: Otto Harrossowitz. 80-89.

2157   ———. 1925. "Papiermuseum und Werkarchiv." *Wochenblatt für Papierfabrikation* 56 (December 19): 1567.

2158   Rohrbach, Kurt. 1974. "Papiergeschichte im Deutschen Museum in München." *Papiergeschichte* 24 (November): 1-6.

2159   Schulte, Ulman. 1959. "Gründung eines Papiermuseums in Spanien." *Papiergeschichte* 9 (April): 34-37.

2160   Süssenguth, D. 1935. "Die Abteilung Papierfabrikation im Deutschen Museum in München." *Der Papier-Fabrikant* 33 (December 8): 585-586.

2161   Tschudin, Peter. 1957. "The Paper Museum at Basal, Switzerland." *The Paper Maker* 26 (September): 1-4.

2162   ———. 1963. "Die Papierhistorische Sammlung im Schweizerischen Museum für Volkskunde in Basel." *Librarium* 6 (August): 79-87.

2163   Tschudin, W.F. 1956. "Orientierung Über die Papierhistorische Sammlung im Schweizerischen Museum für Volkskunde in Basel." *Papiergeschichte* 6 (July): 25-30.

2164   Valls i Subirà, Oriol. 1969. "The Paper Mill Museum of Capellades." *The Paper Maker* 38 (June): 31-37.

2165   ———. 1962. "Das Papiermühlen-Museum zu Capellades und Seine Wasserzeichensammlung. Kurze Geschichte des Papiers in Spanien." *Papiergeschichte* 12 (February): 1-6.

2166   Voorn, Henk. 1967. "A Paper Mill and Museum in Holland." *The Paper Maker* 36 (September): 3-19.

2167   Weiss, Wisso. 1957. "Das Deutsche Papiermuseum." *Marginalien* No. 2 (August): 1-7.

2168   Wheelwright, William Bond. 1948. "Das Dard Hunter-Papiermuseum." *Wochenblatt für Papierfabrikation* 76 (August): 195-197.

2169   ———. 1950. "The World's Greatest Museum of Paper." *The Paper Maker* 19 (September): 11-18.

BIBLIOGRAPHIES

2170 Bartsch, F. 1909. "Books on Watermarks." *Journal of the Royal Society of the Arts* 57 (January 29): 205-207.

2171 Blaser, Fritz. 1958. "Schweizerische Papierliteratur 1950-1957." *Papiergeschichte* 8 (November): 66-70.

2172 Hunter, Dard. 1916. *Handmade Paper and Its Watermarks, a Bibliography.* New York: Technical Association of the Pulp and Paper Industry.

2173 Labarre, E.J. 1957. "Bücher Über Wasserzeichen eine Bibliographie." *Philobiblon (Hamburg)* 1 (September): 237-251.

2174 Laursen, A.R. 1934. *A Bibliography of Watermarks.* Ann Arbor, Michigan: Department of Library Science, University of Michigan.

2175 Schulte, Alfred. 1952. "Ergänzende Literatur zu Briquet's Werk Seit 1907." In *The Briquet Album,* edited by E.J. Labarre. Hilversum, The Netherlands: The Paper Publications Society. 60-68.

2176 _____. 1940. "Die Papiergeschichtliche Literatur von Baden." *Wochenblatt für Papierfabrikation* 71 (July 20): 364-365.

2177 _____. 1941. "Die Papiergeschichtliche Literatur von Württemberg." *Wochenblatt für Papierfabrikation* 72 (September 13): 517-518.

2178 _____. 1939. "Verborgene Papiergeschichtliche Literatur Aus Tageszeitungen und Ihren Beilagen." *Wochenblatt für Papierfabrikation* 70 (April 1): 285-286.

2179 Schulte, Toni. 1953. "Literatur Zur Papiergeschichte Bayerns." *Papiergeschichte* 3 (September): 52-57.

2180 _____. 1955. "Papiergeschichtliche Veröffentlichungen von Armin Renker." *Papiergeschichte* 5 (December): 75-79.

2181 _____. 1955. "Papiergeschichtliche Veröffentlichungen von H.H. Bockwitz." *Papiergeschichte* 5 (September): 57-60.

2182 _____. 1955. "Papiergeschichtliche Veröffentlichungen von Harrison Elliott." *Papiergeschichte* 5 (November): 72.

2183 _____. 1956. "Papiergeschichtliche Veröffentlichungen von N.A. Møller-Nicolaisen." *Papiergeschichte* 6 (April): 14.

2184 Weiner, Jack, and Kathleen Mirkes. 1972. *Watermarking.* Appleton, Wisconsin: Bibliographic Series Number 257, Institute of Paper Chemistry.

2185 Weiss, Wisso. 1956. "Papiergeschichtliche Veröffentlichungen von Franz Zuman." *Papiergeschichte* 6 (November): 70-72.

AUTHOR INDEX

(Numbers refer to entry numbers)

Author Index

SUBJECT INDEX

(numbers refer to entry numbers)

Asia, paper history of, 421-424
Australia, paper history of, 425
Austria, paper history of, 537, 538
Belgium, paper history of, 539-542
Bibliographies of paper history, 2170-
2185
Brazil, paper history of, 1888-1899
Canada, paper history of, 1900
Cartridge paper, 56
Ceylon, paper history of, 426
China, paper history of, 427-433, 435-
437, 439-445, 448-452, 454, 455,
458, 459; Chinese ceremonial paper,
438; invention of paper, 453; rural
paper industry, 434; Ts'ai Lun, 446,
447, 456, 457
Czechoslovakia, paper history of, 543-
559, 561, 563, 570-572, 575-602,
604-608, 612-621; eighteenth
century, 560, 562, 564, 603, 609;
seventeenth century, 566, 567, 573,
574, 603, 610; sixteenth century,
565, 568, 603
Denmark, paper history of, 622-624
Egypt, paper history of, 460-462;
papyrus industry, 463
England, paper history of, 628, 629,
632, 638-642, 644, 647, 648, 651,
654, 655, 666, 670, 671, 674, 677,
678, 689, 693, 709, 715, 717, 725,
728-730; Anglo-Dutch relations, 718;
Balston, William, 627; Barnstaple,
692; Baskerville, John, 643, 649;
Basted Paper Mills, 633; Buckinghamshire, 716; Burnaby, Eustace, 650;

Cornwall, 685, 691, 696-698; Devon,
660, 668, 688, 691, 695-698, 722,
723; Dorset, 699, 706; eighteenth
century, 637, 646, 653, 664, 694,
713; esparto grass, 725; Flintshire,
681; Gloucestershire, 701; Hampshire, 702; Herefordshire, 703;
Hutton's Paper Mill, 680, 682; John
Dickinson and Company, Ltd., 645,
724; Kent, 686; Koops, Matthias,
665, 720, 726; Lindauer Paper Mills,
711; Maidstone, 669, 683; Mathew,
John, 719; Monkton Coombe Paper
Mill, 727; Monmouthshire, 704;
music paper, 667; Nash, William, 684;
nineteenth century, 700; paper
duties, 675; Richmond Paper Mill,
679; seventeenth century, 637, 652,
653, 661-663, 694, 713; Shropshire,
672, 673, 687; sixteenth century,
634, 636, 656, 659, 661, 694;
Somerset, 690, 705, 706; Spielmann,
Johann, 630, 631, 658; Spilman,
Hans, 712; Springfield Paper Mill,
683; Strahan, William, 657; Sussex,
707; Twelve By Eight Mill, 676; wallpaper, 714; Whatman, James, 625,
626, 649, 683, 721; water paper
mills, 710; Wolvercote Mill, 635;
Worcestershire, 708
Finland, paper history of, 731, 733, 734;
first paper mill, 732
France, paper history of, 735, 738-740,
742, 745-748, 761, 768, 778, 780,
781, 788, 789, 794, 795, 798, 799,

155